The Aesthetics of Digital Montage

The Aesthetics of Digital Montage

Film Editing and Technological Change

Marc Furstenau

Routledge
Taylor & Francis Group
LONDON AND NEW YORK

First published in 2024 by Amsterdam University Press Ltd.

Published 2025 by Routledge
4 Park Square, Milton Park, Abingdon, Oxon OX14 4RN
605 Third Avenue, New York, NY 10158

Routledge is an imprint of the Taylor & Francis Group, an informa business

ISBN: 9789463722803 (hbk)
ISBN: 9781041187134 (pbk)
ISBN: 9781003705093 (ebk)
NUR 670

Cover illustration: Elizaveta Svilova in the editing room, in a scene from *Man with a Movie Camera* (Dziga Vertov, 1929).

Cover design: Coördesign, Leiden

DOI: 10.5117/9789463722803

Table of Contents

List of Illustrations

Introduction

Abstract:
In this introductory chapter, I present the basic themes of the book, which is an account of the effects of technological change in the cinema, specifically the changes to the technical apparatus of editing, the advent of digital editing systems. This calls for a new aesthetics of digital montage, I argue, describing what these changes mean for editing as a technique and for the cinema as an art, based on a more careful explanation of the relation between the technical means available to the film artist and the works created through those means, a more coherent theory of the cinematic work of art.

Keywords: Film editing, digital cinema, style, aesthetics, technology, technique

Noël Burch once said, with some considerable theoretical modesty, that "editing as a plastic art is so complex a subject that those of us concerned with film probably do not yet have the means with which to undertake a serious analysis of it."[1] It has arguably become even more complex, as the basic technical apparatus for film editing has been comprehensively transformed, expanding its power as a "plastic art," with significant consequences for our understanding of it as a technique, one so integral to the cinema. What has happened can be described quite simply. The traditionally manual process of cutting and splicing film has been replaced by computerized systems for the organization and manipulation of images and sounds as digital information. While the change itself is easy to describe, the consequences are more difficult to see and understand. Any serious analysis of editing, though, will now have to take this transformation into account, to explain

1 Noël Burch, *Theory of Film Practice*, trans. Helen R. Lane (London: Secker and Warburg, 1973), 47.

Furstenau, M., *The Aesthetics of Digital Montage: Film Editing and Technological Change.* Taylor & Francis Group, 2024
DOI 10.5117/9789463722803_INTRO

what it means for the cinema as an art that editing has been almost entirely reinvented, changed in virtually every aspect as a technical craft.

These changes are a part of the more general transformation of the cinema in the computer age, the comprehensive technological reconfiguration of the traditional apparatus of filmmaking, which has become "digital," supplanting, it is commonly said, its older, original form, which is now called "analog." The process is typically described as a kind of substitution. "The cinema screen," Anne Friedberg wrote so famously, "has been replaced by its digital other, the computer screen."[2] In the digital era, the cinema has lost what she describes as its "medium-based specificity."[3] As Friedberg says, "the chemically-based 'analog' images of photography have been displaced by computer-enhanced digital images."[4] With the loss of such "specificity," with the emergence of a new kind of image, the basic nature or identity of the cinema as an art is supposed to have been changed, paradoxically, it might seem, by expanding the creative powers of the film artist. The new "digital" medium is much more susceptible to modification and manipulation, no longer constrained by what had been understood as significant material limitations. These limitations, though, many argue, are what had importantly determined the formal and aesthetic possibilities of the cinema. They were the source of its artistic value, its identity as an art, what had made the cinema unique.

These limitations have been overcome. Cinematic representations may now be easily "enhanced" by computers, destroying their unique identity. There is an increased ability to alter the basic raw material, now digital rather than photographic. There is a greater creative power of image-making. Robert Stam, for example, has described the "dominance of digital image production," which means, as he says, that "virtually any image becomes possible."[5] The implication, of course, is that the range of possibility had been

2 Anne Friedberg, "The End of Cinema: Multimedia and Technological Change," in Christine Gledhill and Linda Williams, eds., *Reinventing Film Studies* (London: Arnold, 2000), 439. Friedberg mainly emphasizes the effects on "technologies of reception and display," but notes the important changes to "technologies of production," so that, as she says, now "film is commonly edited on video; video is transferred to film; computer graphics and computer-generated animation are used routinely in both film and television production." Ibid. I included this essay in a section on technological change in the cinema in my anthology of classic essays in film theory. See Marc Furstenau, ed., *The Film Theory Reader: Debates and Arguments* (London and New York: Routledge, 2010), 270–81.

3 Friedberg, "End of Cinema," 439.

4 Ibid., 440.

5 Robert Stam, *Film Theory: An Introduction* (Malden, MA: Blackwell, 2000), 319. This is from a section of the book entitled "Post-Cinema: Digital Theory and the New Media," with the

limited before. This limitation is widely thought to have been the source of the cinema's value as an art. Indeed, Stam suggests, like many others, that the changes, while increasing creative power, effectively amount to a kind of aesthetic loss. "The new media blur media specificity," he says. "Digital imaging," as he goes on to argue, "also leads to the de-ontologization of the Bazinian image."[6] It is the loss, that is, so prized by André Bazin, of the supposedly *photographic* identity of the cinema, derived from the raw material of the photographic image, which had been resistant to subsequent alteration. The digital image, by contrast, has a different "ontology," derived from a different material basis, susceptible to unrestrained manipulation and modification. It is merely *simulated*, so easily altered. It has what Stam describes as a "simulacral advantage," but, he adds, there "is also a disadvantage; since we know that images can be created electronically, we are more skeptical about the image's truth value."[7]

It is not clear to me what the "truth value" of an image has to do with the kind of fictional depictions that are characteristic of most film art, but it is an equation that is commonly established, as the basis for claims about the fundamental *aesthetic* transformation of the cinema. In any case, it is indisputable that much has changed *technologically* about film-making. "Digital imaging technologies," noted Stephen Prince in 1996, "are rapidly transforming nearly all phases of contemporary film production. Film-makers today storyboard, shoot, and edit their films in conjunction with the computer manipulation of images."[8] Of these, it is the changes to *editing*—the advent of digital montage, as the specific site for such "computer manipulation," the particular means of digital enhancement, providing the tools for the now virtually limitless mode of image production—that are typically emphasized, where the aesthetic consequences are manifested. It is the new editing technology that is understood, in the most direct sense, to have transformed the basic identity of the cinema. Stam notes that it is the combination of "the digital camera and digital editing (AVID)" that is significant, which will, first of all, "open up montage possibilities," but which will also, and more generally, "inevitably generate new forms of audiovisual intertextuality," an "electronic or virtual textuality," that will in turn "open up intriguing possibilities for both film and film

implication, of course, that the new technology has engendered a new form, fundamentally different from what we had known as "cinema."

6 Ibid.

7 Ibid, 319–20.

8 Stephen Prince, "True Lies: Perceptual Realism, Digital Images, and Film Theory," *Film Quarterly* 49, no. 3 (Spring, 1996): 27.

theory."[9] Nicholas Rombes has asked what it means for the cinema now that "the source material is digitised for editing and becomes part of a digital database," noting that "what the digital makes possible is the absolute disruption of moving images, the very basis of the cinema itself."[10] The disruption is thought to occur at a basic level. Aylish Wood has argued that, traditionally, "through the conventions of editing, space changes from shot to shot through a system of exchange, as one space is replaced by another, and so on."[11] New digital editing technology, though, she says, provides the means for an extensive form of what she calls "morphing," with profound consequences, so that "space can change *within* the framing of the shot."[12] Lucy Fischer, recounting the effects of the new editing technology, makes a similar point, arguing that the digital cinema can now be described as a "virtual cinema—one that questions the 'integrity' of the shot."[13] If, she argues, "film editing meant plastic relationships between shots," now, in the digital era, "editing equally implies synthetic relations *within* shots."[14] As a result, she says, "a new philosophy or ontology of the medium may be in order."[15]

It is a common claim. The new digital editing systems have altered the cinema in some basic, essential way, affecting its very "ontology." The technological changes to the basic tools of film editing are typically thought to be the most significant part of the more general transformation of the cinema in the digital era. In its original analog form, the argument goes, the cinema was "photographic." It consisted, that is, of effectively fixed, unalterable blocks of raw, photographic material, of images or *shots*, characterised by a quality that has come to be described as "indexical," by a supposedly inherent and natural link to or connection with "reality."[16] The "index" has become

9 Stam, *Film Theory*, 322, 324, 326. Stam is referring to one of the first dedicated digital editing hardware systems, created by Avid Technology, Inc., the first version of which was demonstrated in 1988, as I note later, when I trace the history of the new editing technology.

10 Nicholas Rombes, *Cinema in the Digital Age* (London: Wallflower Press, 2009), 73, 76.

11 Aylish Wood, *Digital Encounters* (Abingdon: Routledge, 2007), 50.

12 Ibid., 51.

13 Lucy Fischer, "Film Editing," in *A Companion to Film Theory*, eds. Toby Miller and Robert Stam (Malden, MA and Oxford: Blackwell, 1999), 81.

14 Ibid.

15 Ibid.

16 The term, which has become ubiquitous in film theory, is derived from the philosophical semiotics of Charles Peirce, as an aspect, along with the "iconic" and the "symbolic," of the process that he calls "semiosis." Peirce says, quite simply, that an "*index* represents an object by virtue of its connection with it. It makes no difference whether the connection is natural, or artificial, or merely mental." See Charles S. Peirce, *Collected Papers of Charles Sanders Peirce*, Vol. 8, ed. Arthur W. Burks (Cambridge, MA: Harvard University Press, 1958), 8.368 (n. 23). The

a shorthand term to describe the source of the cinema's artistic identity, in its original form, which is thought to have been lost in the digital era. "Cinema," as Lev Manovich so famously put it, "is the art of the index; it is an attempt to make art out of a footprint."[17] This is changed, though, he argues, by new digital technology, which "subordinates the photographic and the cinematic to the painterly and the graphic, destroying cinema's identity as a media art."[18] For Manovich, and others, this destruction is primarily the effect of new digital editing technology, the emergence of an expanded form of "spatial" montage. The new tools of editing offer an unprecedented degree of control, an increased power of alteration and manipulation. In its original, indexical form, it is claimed, the cinema established the basic limits of editing, determining its value as a technique. Shots could be arranged, placed in a linear sequence, to create all manner of relations and juxtapositions, with various and powerful aesthetic effects, but the shots themselves, the elements within the space of the shot, could not, or not easily, be modified, and therefore *should* not. This is what has changed, as the limits have been removed, precipitating what many describe as an aesthetic crisis for the cinema, as the power of editing has been enlarged and extended, so that, at any point in the filmmaking process, virtually any part of the cinematic representation can be altered—anything goes.

The technical changes to editing have certainly had significant practical effects on filmmaking, upsetting many of the most familiar patterns of working, disrupting the traditional arrangements of the various crafts, altering especially the relationship between the director and the editor, and confusing the distinction between, for example, editing and special effects. What they mean *aesthetically* is less clear, though. There is a long history of technological change in the cinema, and the changes to editing are only the latest, and it is not obvious why they, and not others, alter the cinema's artistic identity or ontology. Indeed, the question of the relation between the physical material of the arts and their value as art has been a fraught one. It

photographic image, though, has been understood by most film theorists as "indexical" just insofar as it has a "natural" connection with the object(s) represented. Peter Wollen may well have been the first to apply the term to cinematic representation specifically, limiting it to a natural connection. "An index," he says, "is a sign by virtue of an existential bond between itself and its object." See Peter Wollen, *Signs and Meaning in the Cinema*, enlarged edition (Bloomington: Indiana University Press, 1972), 122. For an early critique of Wollen's (and others') use of Peirce, see Gorham A. Kindem, "Peirce's Semiotic Phenomenalism and Film," *Quarterly Review of Film Studies* 4, no. 1 (Winter, 1979): 61–69.

17 Lev Manovich, *The Language of New Media* (Cambridge, MA: MIT Press, 2001), 295.

18 Ibid.

has always been difficult to establish any direct link between a particular technical or material change and a formal or stylistic effect, or to say just what the "medium" of the cinema is, what it consists of in any specific or fundamental sense, what material its identity is derived from, and what any technological change might mean for the cinema *as* a medium, *as* an art. Film theorists want to say that the "specificity" or the "identity" of the cinematic medium has been lost with the shift to digital technology. But this is to mistakenly conflate medium and meaning, to confuse physical material with aesthetic value, as John Dewey long ago noted, in light of debates about the then new media of photography and film. "What makes a material a medium," he said, "is that it is used to express a meaning which is other than that which it is in virtue of its bare existence: the meaning is not what it physically is, but of what it expresses."[19]

In this respect, the "expressive" use of any medium in the name of some specific art, is a historical phenomenon, the effect of the choices that artists make, in always changing circumstance, rather than a matter determined by any particular material or technological configuration, or by any particular change to it. Dewey is trying to describe what he calls the "common substance of the arts," distinct from the specific material or materials from which any individual works of art may be rendered. "Media are different in the different arts," he says. "But possession of a medium belongs to them all. Otherwise they would not be expressive, nor without this common substance could they possess form."[20] No specific substance is unique to nor definitive of any individual art, he insists. "It is possible," though, as he says, "for both perceiver and artist to carry their predilection for a particular method of attaining individualization to such a point that they confuse the method with the end, and deny the latter exists when they are repelled by the means used to achieve it."[21] There is just such a "predilection" among

19 John Dewey, *Art as Experience* (New York: Penguin, 2005 [1934]), 209.

20 Ibid.

21 Ibid., 213–14. Dewey was moved to make such an argument partly in response to those at the time who refused to see aesthetic or "expressive" value in the new medium of photography, and the new art of "the movie," which, for Dewey, share an "esthetic quality" with the traditional arts. See ibid., 4. Another American philosopher, D. W. Prall, was making a similar argument at around the same time, rejecting any *a priori* distinction between media as instruments of aesthetic expression, arguing for the artistic value of photography and film. "There is nothing intrinsically spiritual," Prall says, "in charcoal or paint-brushes, palettes and stretched canvases, red-lead and ochre and oil, to make them superior to lenses and shutters and gelatin-coated films. And there is no reason," he adds, "in the nature of things why subjects before a camera may not have expended upon them the whole power and skill of a great artist, or why, for his preservation of his vision of them, a camera of sufficient delicacy and contrivance might not

film theorists for a "particular method" in the cinema, for a specifically photographic materiality, which had initially "repelled" defenders of the traditional arts, in Dewey's time, against whom he is arguing, but which is now championed, thought to be the source of a unique aesthetic quality. There is, though, still a confusion of means and ends, and still a tendency to distinguish photography and film from the other arts, and to make the corollary claim that that distinction has been destroyed.

To acknowledge that there has been a significant technological change in the cinema, that editing especially has been thoroughly, indeed comprehensively, transformed, does not mean that the very nature of the cinema, its artistic identity or ontology, described as photographic, or "indexical," as "analog," has been changed or destroyed, replaced by a wholly new material basis or medium, which is electronic, "symbolic" or "digital," with different or even conflicting aesthetic value. There is, as Berys Gaut has argued, an important distinction to be made between the technical and the aesthetic, between a "medium" as the material substance from which a work of art may be fashioned, and the actual work of such fashioning. He insists that, "though digital cinema is a new medium, it does not follow that it is a new *artistic* medium."[22] David Davies has similarly argued that "a loss of indexicality consequent upon the use of digital technology does not in itself justify the claim that digital cinema differs fundamentally from its traditional counterpart."[23]

Both are making a principled distinction between the technical and material facts of filmmaking, on the one hand, and the cinema as an art, on the other, between the material substance of film art and the formal and artistic act of rendering that substance, of forming that material, a

be the best means." See D. W. Prall, *Aesthetic Judgment* (New York: Thomas Y. Crowell Co., 1929), 198–99. More recently, and in the same vein, Irving Singer has said that most film theorists place too much emphasis on the material aspects of the technical apparatus of photography, making the same basic error as those who had originally opposed the new instruments, assuming that the value of cinematic imagery derives directly from the material nature of the apparatus, from the technical processes of photographic creation. Singer argues, though, that cameras are effectively no different than any other device, that photography and film are not unique media, but can, like any other medium, be put to use by artists to various aesthetic ends. "By examining the technical details of film," he says, "we learn how its transformations are produced, and how they affect the audiences they can reach. The images made and organized by the camera," though, he says, significantly, "are of little interest in themselves or in their formal structure." See Irving Singer, *Reality Transformed: Film as Meaning and Technique* (Cambridge, MA: MIT Press, 1998), 11.

22 Berys Gaut, "Digital Cinema," in *The Routledge Companion to Philosophy and Film*, eds. Paisley Livingstone and Carl Plantinga (Abingdon: Routledge, 2009), 78.

23 David Davies, "Digital Technology, Indexicality, and the Cinema," *Rivista di estetica* 46 (2011): 58.

distinction that I will also try to make. No straightforward relation can be established, I will argue, between any particular technological change and the aesthetic identity of the cinema. Whatever industrial, economic, and financial disruptions might result when such changes occur, the cinema itself, the cinema as an art, as a mass medium, as a modern entertainment, generating fictions, creating imaginative representations, persists, even as significant parts of the technical apparatus of filmmaking are altered. For many film theorists, though, the most recent changes to the cinema, and especially those to the technical apparatus for editing, seem to pose a profound threat. The very identity of the cinema, its "ontology," its "indexical" identity, seems to have been seriously affected, if not in fact destroyed. A kind of category change is supposed to have occurred, with the original cinema replaced by some other kind of cinema—or by some new phenomenon that cannot even be described any longer as cinema, as it has traditionally been understood.

The supposed crisis of the cinema, though, the crisis of the digital cinema, is better understood as fundamentally a discursive crisis, as Stephen Prince in fact noted, describing the initial stages of the technological transformation of the cinema. "The rapid nature of these changes," he said, "is creating problems for film theory."[24] Given the novelty of "the digital manipulation of images," the "unprecedented" nature of the "creative possibilities it offers," he says, "its effects on cinematic representation and the viewer's response are poorly understood. Film theory has not yet come to terms with these issues."[25] It has still not fully come to terms with them, I will argue, especially with the issue of editing. A significant part of the problem is that, while the changes to editing are thought to be the primary source of the transformation of the cinema as an art, new editing systems understood as the means by which such digital manipulation is made possible, editing itself, as a technique, still largely resists easy explanation. The traditional analyses of editing, to the degree that they have been able to explain the technique (or, perhaps, as the reason they have failed to explain it, as Burch suggests), have depended on certain untenable assumptions about the cinema, about the nature of the shot, of cinematic representation or depiction, about the supposed "ontology" of the photographic image, and about the basic relation between technology and art, between the technical and the aesthetic, between the physical material used in the fabrication of a work of art and the intentional and creative activity of the film artist.

24 Prince, "True Lies," 27.

25 Ibid.

The advent of a digital cinema, and of new digital editing systems, the emergence of a new form of digital montage, can be understood to provide, in the simplest sense, the technical means for easily altering elements within the shot, expanding the power of editing beyond the linear arrangement of shots to include the manipulation and alteration of the content of the shot. This may be accounted for, though, according to what can be described as an intentionalist aesthetics, as proposed, for example, by Gérard Genette, who offers a "theory of the artwork," which must, he says, "take into account at the same time, as much as possible, the work as *object* and the work as *action*."[26] In much film theory, though, and especially in accounts of the digital cinema, and of digital editing, such a distinction is elided. Intentional action is typically subordinated to the material constitution of the object. Artistic creativity is understood as the less significant source of aesthetic value in the cinema, which is supposed to derive instead primarily from the "medium." Indeed, new digital editing technology is typically understood as an unwarranted and worrying *expansion* of intentional, artistic control, overcoming the kinds of physical constraints that had been (properly) imposed on the film artist by the particular nature of the filmic object, given the basic, physical resistance of the cinema's raw material, supposed to be the photographic image. It is a curious, even paradoxical, part of much of the theory of the digital cinema, that the affordances and capabilities offered by new digital technology, and especially by the new editing software, are understood to have diminished the cinema as an art, altering its original identity, even destroying what had in fact distinguished it, making it no different now than any other art. While there are, of course, effects that arise from any technological change, that can make life better or worse for an artist, that can significantly alter their working conditions, providing

26 Gérard Genette, "From Text to Work," in *Essays in Aesthetics*, trans. Dorrit Cohn (Lincoln: University of Nebraska Press, 2005), 26; emphases in original. Dewey, Prall, and Singer, are also effectively emphasising the fact of intention in art. Others have offered what can be called an intentionalist aesthetics, in several works that generally and implicitly inform my argument. Gregory Currie, for example argues that "art works are in some sense closely connected with human action," and endeavours, as he says, to make this "widely recognised" fact more theoretically precise. See Gregory Currie, *An Ontology of Art* (London: Macmillan Press, 1989), 1. Contending with the undeniable complexities of the question of intention in art, David Davies offers a theory of the work of art as "performance." See David Davies, *Art as Performance* (Malden, MA: Blackwell, 2004). Berys Gaut and Paisley Livingston have edited a collection of essays that emphasize the fact of creativity, the "creation" of art, that seek to redeem the notion of intention in analyses of art. "Works of art," they say, in their introduction, "unlike natural objects, are after all *works*, the products of makings." See Berys Gaut and Paisley Livingston, eds., *The Creation of Art: New Essays in Philosophical Aesthetics* Cambridge: Cambridge University Press, 2003), 2.

new possibilities but also imposing new kinds of constraints, there is a kind of logical error in the claim that any particular change necessarily alters the essence or identity of an art, that a new and distinct "digital" cinema has emerged, offering greater intentional power, replacing the cinema in its traditional form, which had been understood as necessarily limited in its intentionality.

There is now a vast literature on the subject of the "digital cinema," which for the most part makes this kind of claim.[27] The shift to the digital is usually presented as an essential change, in very general terms, with little attention paid to the much more specific or local changes, to particular tools, as distinct parts of the technical apparatus of filmmaking. In most accounts, the question of editing is inevitably raised, but discussion of it is necessarily subordinated to the more general claim. The changes to editing are thought to be where the formal and aesthetic effects of new digital technology are manifested most profoundly, with such enormous consequences for the cinema's very identity as an art, but editing itself in the digital era is rarely discussed in any detail. An account of the changes to editing, though, can be offered that describes the very real effects of these, but does not understand them to have altered the cinema in any essential way. Editing, in both its original form, and as it has been technologically transformed, is best understood as an aspect of the cinema as a creative art, characterized generally by intentional, imaginative manipulation, artistic transformation, by the marshalling of the various parts of the technical apparatus of filmmaking, to whatever chosen extent, to aesthetic ends. That apparatus is always subject, though, to all manner of technological change. Indeed, the history of the cinema is largely a history of technical transformation. The changes to editing, while significant, are only a particular instance of the almost constant technical remaking of the cinema, of the always changing

27 Increased intentional control is implied as the primary effect of new digital tools when, for example, Philip Hayward and Tana Wollen say that, in the digital era, "audiovisual technologies have apparently removed barriers between the real and the represented, … and have ironically rendered problematic indexical relationships" between real objects and their representation. Philip Hayward and Tana Wollen, eds., *Future Visions: New Technologies of the Screen* (London: BFI, 1993), 2. Some other notable works, making similar claims, are: Thomas Elsaesser and Kay Hoffmann, eds., *Cinema Futures: Cain, Abel, or Cable? Cinema Futures in the Digital Age* (Amsterdam: Amsterdam University press, 1998); Matt Hanson, *The End of Celluloid: Film Futures in the Digital Age* (Mies: Rotovision, 2004); Stephen Keane, *CineTech: Film, Convergence, and New Media* (Basingstoke: Palgrave Macmillan, 2007); Laura Mulvey, *Death 24x a Second* (London: Reaktion, 2006); Jeffrey Shaw and Peter Wiebel, eds., *Future Cinema: The Cinematic Imaginary After Film* (Cambridge, MA: MIT Press, 2003); Garrett Stewart, *Framed Time: Toward a Postfilmic Cinema* (Chicago: University of Chicago Press, 2007).

technological conditions in which filmmakers undertake through their actions to render objects of aesthetic interest, to create works of film art.

* * *

The change from film editing to digital montage is now a part of the history of cinematic art. The tools have changed. The technical apparatus has been thoroughly transformed. The basic scene of editing, though, as an aesthetic undertaking, as artistic action, is effectively no different. Almost a hundred years ago, Elizaveta Svilova sat in an editing room in Moscow, carefully arranging the shots in one of the most famous films about editing, *Man with a Movie Camera* (1929), directed by her husband, Dziga Vertov. It is a still image from the film, showing Svilova at work, that is on the cover of this book, and which I discuss in more detail later. In a self-reflexive gesture, the editing of the film is incorporated into the film itself, as we see Svilova cutting and splicing strips of film, to produce the avant-garde, montage effects that are on display, for which Vertov, and the early Soviet filmmakers more generally, were to become so famous. Just as important as, or even more important than, the man with the movie camera, it seems, is the woman with scissors and glue. What Svilova is doing with such simple tools, with the strips of film arranged on the light-board before her, like an illuminated screen, so similar to the computer monitors and software interfaces of modern digital editing systems, is comparable, in basic aesthetic terms, to the work of editing today. The aesthetics of digital montage, the formal and artistic possibilities that emerge from the manipulation and arrangement of the various elements of cinematic representation, made possible by the new electronic and computer technology, derive from the art of film editing as it developed through the efforts of such figures as Svilova and Vertov.

There is a story to be told that links the scene of Svilova cutting and arranging strips of film to the work of editors in digital editing suites. It is a story, as it is for any art, of the relation between the tools used by the artist, the material conditions of their artistry, and the realization of their artistic goals, the creation of a work of art—the relation, more fundamentally, between the technical and the aesthetic. This relation, as I will argue, is manifested in the familiar distinction between "editing," on the one hand, the specific physical and technical craft, and "montage," on the other, the concepts guiding such work, a theory or set of aesthetic principles, like those that were being developed in the work of Svilova and Vertov. They were part of a large and heterogenous group of filmmakers and theorists, working in the early Soviet era, from whom we inherit the idea of *montage*, an explicitly aesthetic idea of editing. The historical and conceptual continuity that I will argue for is manifested in the phrase "digital montage." If there is an "aesthetics" of editing in the computer era, it is part of a more general cinematic aesthetics, developed historically, related to and derived from these sorts of early efforts, by these film artists, who were conceiving of editing as an artistic practice, just as *montage*.

Still, despite such conceptual or aesthetic continuity, there is indeed much about editing that has changed. I will be describing the changes to the tools of film editing, to the basic technology of editing, and considering the effects of these on editing as a craft. The advent of fully functional digital editing systems, as Leo Enticknap has argued, depended crucially and specifically on the development of microprocessor technology, and the increasing ability to effectively manage large amounts of audio-visual information in the form of digital files. While the microprocessor, the "integrated circuit," or the "microchip," was invented, as Enticknap notes, in 1958, it was not really until about the end of the 1990s or the early 2000s, that "video manipulation and editing would be possible using home computers of average power and memory capacity."[28] Various specific technical constraints limited the early development of digital editing. "Once microprocessors were fast enough," though, Enticknap says, "it made obvious sense for videotape recording to go digital."[29] Once video had gone digital, it was available for a new comprehensive form of editing—digital montage. This subsequent development was just as obvious. The new affordances of digital video were attractive, and they could be fully realized with the right tools. Dedicated digital editing systems were invented, first requiring specialized hardware, elaborate suites of devices, but soon just stand-alone software programs, that could be run on ordinary personal computers. The history of the cinema is now fully integrated, so to speak, with the history of computing.[30]

28 Leo Enticknap, *Moving Image Technology: From Zoetrope to Digital* (London: Wallflower Press, 2005), 206. For more details on the invention and the development of the integrated circuit, or the microprocessor, see Martin Campbell-Kelly and William Asprey, *Computer: A History of the Information Machine* (New York: Basic Books, 1996) and Paul Ceruzzi, *A History of Modern Computing* (Cambridge, MA: MIT Press, 1999).

29 Ibid., 210.

30 There is, of course, another way of thinking about this "integration," about the relation between the cinema, understood as a supposedly "photographic" medium, and the history of computing, that does not see them at odds, one replacing the other, but rather acknowledges the much more complex relations between the technologies, specifically the role of photography in the invention of the microprocessor, the microchip, itself, as Patrick Maynard has noted. "If something called photography," says Maynard, "seemed about to be transformed by fast-developing computer technologies, these technologies were themselves functions of a photographic technology and

Enticknap and others provide detailed accounts of this technical history. I am more concerned, though, with the *aesthetic* effects of new digital editing technology, with the consequences of these for our theoretical understanding of the cinema as an art. The changes are effectively forcing a conceptual reconfiguration. If the crisis of the digital cinema is, in fact, as I say, primarily a discursive crisis, what is required, to a large degree, is a new theoretical vocabulary, or, as I will be suggesting, a broadening of our vocabulary, of our conceptual framework. If, as Burch insists, we have not had the means with which to undertake a serious analysis of editing, this is partly the result, I contend, of certain persistent conceptual confusions at the heart of film theory, according to which particular technical changes, like those to editing, tend to be exaggerated in their effects on the cinema as an art. An effort to clarify our concepts, as I will try to show, will help us to understand what has indeed happened to editing, as a technology, but without positing some kind of ontological break in the history of the cinema, without assuming any particular aesthetic consequences.

This is not to diminish the scope of the change, though. The traditional tools of film editing have been replaced by an entirely new technical apparatus. No longer a physical process of cutting and splicing strips of celluloid, film editing has been transformed into an extensive process of digital montage, the manipulation and modification of practically all aspects of the image and sound. The new technology offers an unprecedented degree of fine-grained control over virtually every element of cinematic representation. This is undeniably significant. In the most important respect, though, the fundamentals of editing remain the same. With complex computer software, with the new digital editing systems, editors and filmmakers are able to arrange shots much more efficiently, but the patterns of arrangement remain, for the most part, quite familiar. The ultimate effect is still *linear*, with films consisting of a series of shots, one after the other, organized in some kind of coherent and meaningful order. The technology itself, though, the editing software, is importantly *nonlinear*, allowing for effectively random access to any part of the recorded material, rather than having to scroll through lengths of film, cutting out pieces and gluing them back

could not exist without it. The reason is simple: personal computers were made possible by the emergence in the 1970s of microprocessors, tiny solid-state structures that depend, crucially, upon the process of microfabrication of computer chips, which is itself a 'photolithographic' technology." See Patrick Maynard, *The Engine of Visualization: Thinking Through Photography* (Ithaca and London: Cornell University Press, 1997), 6.

together, as in the past.[31] This is also what extends the power of editing, though, so that it is no longer only a matter of arranging shots in order, placing relatively fixed images in a linear sequence. It is now possible to "edit" within the shot itself, after the fact of recording, to more easily manipulate and alter the content of the image, which is now rendered as electronic or digital information, allowing access to every part of the representation, at the more minute level of what is often described as the *pixel*, a neologism, a contraction of the phrase "picture-element." If it had once been the "shot" that was understood as the smallest basic element of cinematic composition and representation, the scale has been reduced, to the pixel, to an effectively atomic level, making every part of the representation subject to subsequent manipulation and alteration.

There are important consequences as a result of such changes, and film artists have taken advantage of the new technical affordances. Rather than describing these, though, as instances of artistry, understood precisely as the creative use of the various tools and materials available to a filmmaker, broad generalizations about the effects on film as an art are typically offered. A wholly new theory of the *digital* cinema is supposed to be needed, as an art now characterized by an unprecedented degree of intentional control, of imaginative power, provided in the main by the new tools of editing. A new kind of cinema seems to have emerged, which is, ironically it seems, and in this respect, now no different than any other art—as "subjective" as any other art, having lost the essentially "objective" character of the photographic cinema. Gone is what seemed to have made the cinema aesthetically distinctive, unique among the arts, namely an inherent and important material constraint on intention. The technical advances in cinematic technology are thought, in this respect, to have had a kind of retrograde or diminishing effect on film art. Given the technical transformation of editing specifically, with the new tools for manipulation and alteration available to the filmmaker, the traditional constraints in the cinema have been removed, constraints that had been understood to define the cinema as an art. There are, many claim, no longer any limits on the filmmaker's imagination, no longer any boundaries to their intentional control.

31 Nicholas Rombes considers the ambiguous relation between the ideas of the "linear" and "nonlinear" in digital film editing, as a practical craft, and in film as an art. "Symbolically," he says, "nonlinear editing corresponds to traditional film editing that took place on uprights and flatbeds: in both cases, images are managed and rearranged to tell a story. But the random access and virtually instant recall of the of the database fundamentally alters the way we perceive of the relationship between the shots in a film." Rombes, *Cinema in the Digital Age*, 73. This last claim, though, is one that I will challenge.

What characterizes the cinema as an art, though, one could argue, is not any particular technical apparatus, any specific tool, any unique material basis, but rather, indeed, the very fact of intention, the deliberate use, that is, of any technology, tool, or medium, any material or substance, put to imaginative or aesthetic ends, made to create an object, a *work of art*, that performs an aesthetic function. The history of art, of any art, is to a large and important extent a history of technological change. The cinema is no different. Many of the material aspects of art change, but what Dewey described as the "common substance" of art remains the same. The most recent changes to the cinema, in this respect, while extensive, while significant, are commensurate with an account of the cinema as an art. The changes to editing are, arguably, the most extensive, the most significant, but they can be explained without assuming that the cinema, *as an art*, has changed. Still, given the importance of editing as a technique, given that editing has long been understood as a technique unique to the cinema, and particularly important to the artistic identity of the cinema, it is worth carefully considering the change to the basic technology of editing, the transformation from film editing to digital montage.

* * *

The history of the advent of digital editing can be briefly summarized.[32] In the most basic sense, as I have said, it begins with the invention of the integrated circuit, the microchip, the power of which can only be applied, though, once the basic material of the cinema is transformed. That transformation arguably begins with the invention of the Ampex electronic video recorder in 1956, with the advent of videotape, as an alternative to film, and then with the introduction of Editec, in 1962, the first dedicated electronic editing system, with which the new video recordings could be effectively and efficiently edited. By 1970, on the basis of this initial technical development,

32 What I offer here as a summary is based primarily on Michael Rubin, *Nonlinear: A Field Guide to Digital Video and Film Editing*, 4th ed. (Gainesville, FL: Triad Publishing, 2000), the best and most detailed account of the transformation, presented as a year-by-year chronicle, with a wealth of technical and corporate detail. Rubin published the first edition of his book in 1991, revising it in 1992, and 1995, and finally in 2000, in light of the many rapid changes taking place in that decade. See also Enticknap, *Moving Image Technology*, for a good summary of the change to digital technology, including editing technology. For broader contextual accounts of the history of the transition to digital editing systems in the American cinema, see Deron Overpick, "The New Hollywood, 1981–1999: Editing," and Meraj Dhir, "The Modern Entertainment Marketplace, 2000–Present: Editing," in Charlie Keil and Kristin Whissel, eds., *Editing and Special/Visual Effects* (New Brunswick, NJ: Rutgers University Press, 2016), 129–41, 156–71.

the first computer editing system was created, the CMX 600—the product of a joint venture between the American television broadcast network CBS and the magnetic recording tape company Memorex. At first, these technologies were being used only in the still very new context of television production. Some of the effects of this technological change were being felt in the film industry, but still only in stylistic terms, as filmmakers were beginning to notice the kind of editing in television programs, which they would sometimes emulate.[33] Editing in the cinema, though, was still almost exclusively *film* editing, still done manually, by cutting strips of celluloid and reassembling the pieces. That changes only in the 1980s, with such systems as the Montage Picture Processor, a nonlinear videotape editor, and the EditDroid system, developed by Lucasfilm, George Lucas' production company in San Francisco. Stanley Kubrick used the Montage system to edit his film *Full Metal Jacket* (1987), as did Alan Alda, directing his first film, *Sweet Liberty* (1986), and Sidney Lumet, on *Power* (1986). Other filmmakers begin using it, too, and, in 1987, acknowledging the significance of the new system, and its growing popularity, the inventors of the Montage Picture Processor were given a special Academy Award for technical development.[34]

By 1989, the first truly digital, nonlinear, electronic video editing systems are invented, and many more filmmakers began using them, most notably the Avid/1 system, and Lucas' redesigned (and renamed) EditDROID, first used by Oliver Stone to edit parts of *The Doors* (1991). The era of digital montage in the cinema had begun. In 1991, Apple released QuickTime, to run digital video on a personal computer, opening the door to the first digital stand-alone editing software programs like Apple's Final Cut and Adobe Premiere, which could be run on an ordinary desktop or laptop

33 The Canadian-born Hollywood director Edward Dmytryk was one of the first to notice what was happening in television, and to speculate about the possible effects on film editing as a result of such stylistic emulation of the televisual aesthetic, by the new video technology. Known for such films as *Farewell My Lovely* (1944), *Crossfire* (1947), and *The Caine Mutiny* (1954), Dmytryk began his career as a film editor in the 1920s and 1930s. At that time, he says, editing had been "severely depressed by the advent of sound, with its 'fixative' tendencies." In the early 1980s, he says, there is a similar technical crisis in editing, noting that "it is once again suffering through a period of banality, which is the inevitable result of the peculiar economic and so-called artistic demands of corporate television." While he does not mention the new video editing technology explicitly, he was keen to address what he nevertheless saw as a significant threat to the technique from the new technology of television, writing his manual in the hope, as he says, of "reawakening an interest on the part of filmmakers in the still unexplored potential of film editing." See Edward Dmytryk, *On Film Editing: An Introduction to the Art of Film Construction* (Boston: Focal Press, 1984), x.

34 Rubin, *Nonlinear*, 59.

computer. It is with the purpose-built, dedicated systems, the digital editing workstation or editing suite, though, that the revolution in film editing is initiated. Describing the changes, Stephen Prince notes that, by the end of the 1980s, "post-production practices were undergoing major redesign. The traditional approaches to film editing," he says, "vested in the physical acts of cutting, splicing, and searching through trim bins had given way to the 'cleaner' and more powerful use of electronic and then digital technology to offer quicker and more flexible approaches."[35]

By the beginning of the 1990s, the changes were being effectively consolidated, as electronic and digital systems became standard. As Michael Rubin observes, with the adoption especially of the Avid systems by many Hollywood studios, "1992 [was] a watershed year for nonlinear editing, and digital nonlinear editing in particular."[36] As Rubin notes, J&R Film, a major technical supply company in Hollywood that rented out editing machines, mainly flatbed systems like the Steenbeck and the K-E-M, began to acquire digital nonlinear systems. Very soon these were the only ones that filmmakers wanted, the Avid system being the most popular. The old devices were phased out. In an especially symbolic move, J&R stopped renting out the Moviola, an older stand-up editing device, a venerable machine, the first dedicated apparatus for editing, that had been introduced in 1924, and which many editors had continued using.[37] "It was a landmark move," Rubin

35 Stephen Prince, *A New Pot of Gold: Hollywood Under the Electronic Rainbow, 1980–1989*. History of the American Cinema, edited by Charles Harpole, vol. 10 (New York: Charles Scribner's Sons, 2000), 115. Prince adds, significantly, that the changes to editing technology blurred or dissolved some of the most basic distinctions between elements of the post-production process. "Editing and special effects work," he says, "bonded intimately as related parts of a unified phase of electronic post-production" (115). Prince pursued the question of special effects in a later book. See Stephen Prince, *Digital Visual Effects in Cinema: The Seduction of Reality* (New Brunswick, NJ: Rutgers University Press, 2012). For further considerations of the relations between editing and special effects, in both the traditional cinema and the later digital era, see the essays in Martin Lefebvre and Marc Furstenau, eds., *Special Effects on the Screen: Faking the View from Méliès to Motion Capture* (Amsterdam: University of Amsterdam Press, 2022).

36 Rubin, *Nonlinear*, 64.

37 The Moviola was invented by Iwan Serrurier, and was in fact patented under that brand name in 1919 as a "Picture Presenting Apparatus," imagined initially as the means by which movies could be shown in private domestic settings, before becoming, with some slight redesigning, an exclusively technical device used for editing, as Eric Theisen explained, in an article in a trade journal in 1935, when it had become the standard device in editing rooms. "The first moviola," he says, "was a far cry from the present device; in fact it was a home movie projector in a victrola-like cabinet." See Eric Theisen, "The Story of the Moviola," *The International Photographer* 7, no. 10 (November 1935): 4; cited in Paul Monticone, "Classical Hollywood, 1928–1946: Editing," in *Editing and Special/Visual Effects*, eds. Charlie Keil and Kristin Whissel (New Brunswick, NJ: Rutgers University Press, 2016), 205, n. 9.

says, "in an industry where the Moviola had been the bedrock of traditional editing equipment for 100 years."[38] A new, digital foundation was being built, replacing the traditional material infrastructure with new kinds of systems.

This brief capsule history provides an initial account of the relatively rapid transformation of film editing to digital montage. I will provide more historical detail as I proceed, but the basic story of the change is relatively straightforward, and the effects on the practical aspects of filmmaking, and even the broader industrial and economic effects, can be easily recounted. The aesthetic consequences of these changes, though, as I say, are more difficult to describe. To do so will require revisiting the history of thinking about editing, revising some of the most basic assumptions about the technique, which have been at the core of much film theory. For many, and according to such assumptions, editing is a basic technique, defining the cinema as an art. Or, more precisely, it is *film* editing that had defined the cinema, determined by the even more basic, supposedly "ontological" fact of photography, understood as the primary source of the "raw material" of the cinema, the photographic image, in the form of the shot. A series of shots could be arranged and rearranged by editing, but they were not easily altered or modified themselves. The shot, as the solid, elemental component of the cinema, determined the very limits of editing as a technique, in turn defining the cinema as an art. Film editing has been replaced, it is argued, by a new and more comprehensive digital process extending or destroying such limits. A technique that had been defined by material limitation has given way to a new process of unlimited manipulation. Yet certain basic aesthetic assumptions must apply whether one is cutting and splicing strips of film or manipulating audio-visual data on a computer. The fact of technical change must be understood as a necessary part of the very conception of art, perhaps especially the art of the cinema. There is a basic relation between technology and art, between the technical and the aesthetic, that I will explore, and that allows for an explanation of technological change as a necessary part of the history of cinema as an art, consisting of such techniques as editing, according to which a continuity may be established between the original technical apparatus of editing and the new digital systems.

Most accounts of editing posit a basic relation between the technical and the aesthetic, but often in too simplistic, even deterministic terms. An important part of the idea of editing, as it has been developed by film theorists, is that the aesthetic effects that can be created, the most complex artistic possibilities, derive from the very basic tools and procedures of

38 Rubin, *Nonlinear*, 65.

editing as a practical craft. In strictly technical terms, editing is (or was) perhaps the simplest of all the techniques of filmmaking, and it can be described very easily. Jean Mitry, for example, has offered a basic definition. "Editing (*montage*)," he says, "technically speaking, is nothing more than the laying end to end of different shots."[39] So simple, but so significant. Indeed, its significance seems to derive just from its simplicity. Editing, this rudimentary procedure, the mere linear arrangement of shots, is of particular importance, in its simplicity, performing (at least potentially) an aesthetically constitutive function, as Mitry suggests. "This work, "he says, "strictly artisanal, is one of the most crucial operations of cinematic art."[40] Indeed, it is only with the invention (or discovery) of editing, many claim, and the aesthetic possibilities that it created, that the cinema emerged *as* an art. It was when filmmakers first had the idea to join separate shots together, as, for example, André Malraux argued, "that the cinema as an art was born."[41] For Béla Balázs, it is editing specifically that creates artistic value in the cinema. "In film," he says, "the most meaningful set-up does not suffice to give the image its full meaning. In the final analysis, meaning is determined by the position of the image between other images. The issue here is editing ... this is the ultimate refinement of work on film."[42] Even André Bazin, who so famously insisted on the "limitations" of montage,

39 Jean Mitry, *The Aesthetics and Psychology of the Cinema*, trans. Christopher King (Bloomington and Indianapolis: Indiana University Press, 1997), 92. The importance of the distinction in English between the practical fact of "editing" and the artistic technique of "montage" is implied by the parenthetical inclusion here in the English translation. In French, Mitry says simply: "Le montage qui, techniquement, n'est rien de plus que la mise bout à bout des différentes prises de vues," adding that editing (*montage*) "est aussi vieux que le cinéma lui-même," that editing "is also as old as the cinema itself," implying that the technique is an original and perhaps essential aspect of film art. See Jean Mitry, *Esthétique et psychologie du cinéma I: Les Structures* (Paris: Éditions Universitaires, 1963), 271. On these grounds, as we shall see, Mitry also establishes a distinction between the practical fact of editing and an aesthetics of montage, building on and revising Soviet-era montage theory.

40 Mitry, *Esthétique et psychologie du cinéma I*, 354; my translation. "Ce travail, strictement artisanal, est l'une des opérations capitales de l'art cinématographique." In the published English translation, this passage is rendered, rather less literally, as: "This purely technical stage in the filmmaking process is of the utmost importance." Mitry, *Aesthetics and Psychology*, 127.

41 Quoted in Mitry, *Aesthetics and Psychology*, 67. From André Malraux, "Esquisse d'une psychologie du cinéma," *Verve* 2, no. 8 (June 1, 1940): 69–73. Translated as André Malraux, "Sketch for a Psychology of the Moving Pictures," in *Reflections on Art: A Source Book of Writings by Artists, Critics, and Philosophers*, ed. Suzanne K. Langer (Baltimore: The Johns Hopkins Press, 1958), 317–27.

42 Béla Balázs, *Béla Balázs: Early Film Theory*, Visible Man *and* The Spirit of Film, ed. Erica Carter, trans. Rodney Livingstone (New York and Oxford: Berghahn Books, 2010), 122. This is a single-volume English translation of Balázs' two books on film. The quotations here are from

was also willing to accept its very real "virtues." While he advocated for a restricted use of editing, he understood and acknowledged its fundamental necessity. It was, for him, "in no sense a question of being obliged to revert to a single-shot sequence or of giving up resourceful ways of expressing things or convenient ways of varying the shots."[43] Siegfried Kracauer, too, similarly ill-disposed to "formalistic" manipulations of the original photographic camera recording, admits to the power and necessity of the subsequent ordering of the shots. "Of all the technical properties of film," he says, "the most general and indispensable is editing."[44]

It was, of course, the early Soviet-era theorists and filmmakers who argued most explicitly and passionately for the values of editing, which they thought performed an elemental and definitive role. "The foundation of film art," declared Vsevolod Pudovkin, "is *editing.*"[45] They established a basic distinction, now a familiar one, between "editing," the word that Pudovkin uses here, understood as the necessary practical and technical procedure of arranging shots, and "montage," imported from French, rendered in Russian as *montazh*, which they defined as the artistic or aesthetically constitutive use of editing. Editing, of course, quite simply and practically, is little more than the physical connection of one shot to another, creating a linear chain or sequence of shots. One needs, though, they argued, a theory of connection, guiding the process, proposing the concept of "montage," understood as the deliberate *assembly* of the finished film from a raw material, or the ultimate *arrangement* of basic elements, which is then conceived of as the fundamental source of meaning or significance in film. "Montage," as Lev Kuleshov insisted, is "the very basis of cinematography."[46] The most

The Spirit of Film, published originally in German as *Der Geist des Films* (Frankfurt: Suhrkamp Verlag, 1930).

43 André Bazin, "The Virtues and Limitations of Montage," *What Is Cinema?*, trans. Hugh Gray (Berkeley: University of California Press, 1971), 50.

44 Siegfried Kracauer, *Theory of Film: The Redemption of Physical Reality* (Princeton: Princeton University Press, 1960), 29. Kracauer distinguishes between such "technical properties," which include the techniques of cinematography, the various "special effects," and so on, editing being the most important of these, and the "basic properties," which are, he argues, "identical with the properties of photography." According to this distinction, he presents his basic thesis: "Film, in other words, is uniquely equipped to record and reveal physical reality and, hence, gravitates toward it." Ibid, 28.

45 V. I. Pudovkin, *Film Technique and Film Acting*, trans. Ivor Montagu (London: Vision Press, 1954), xiii; emphasis in original.

46 Lev Kuleshov, *Kuleshov on Film: Writings of Lev Kuleshov*, trans. and ed. Ronald Levaco (Berkeley, Los Angeles: University of California Press, 1974), 48; emphasis in original. Kuleshov is describing the work that he undertook with his students in Moscow in the 1920s, experimenting with the possibilities of editing, on the basis of which, he says, "we came to understand that the

basic principle of montage theory, as Kuleshov argued, is that editing is to be distinguished from the substance that was being edited. "Separate shots of film," as Kuleshov says, "constitute cinematic material."[47] They await expressive fulfilment, so to speak, through subsequent manipulation. Aesthetic value in the cinema does not reside in the shots, that is, but is derived specifically from their ordering. It is Sergei Eisenstein who develops this claim in its most famous and influential form, producing the most elaborate and intricate theory of editing as montage. "Shot and montage," he says, "are the basic elements of cinema. Montage," he adds, with a telling metaphor, "has been established by the Soviet film as the nerve of cinema."[48] Indeed, the cinema as an art, he insists, is effectively reducible to the fact of editing, to montage. "To determine the nature of montage," he says, "is to solve the specific problem of cinema."[49] Of course, the "essence" of montage cannot be reduced to the basic technical procedure of editing, as he admits. "The mechanical process of splicing

basic strength of cinema lies in montage, because with montage it becomes possible both to break down and to reconstruct, and ultimately to remake the material." Ibid., 52. In the English translation of Kuleshov, the term "montage" is used consistently. A distinction is maintained in the translation of Pudovkin, though, between "editing," used in the passage I quote above, and "montage," which is presented as a distinct term, which he uses later in the book. It was from Kuleshov, Pudovkin says, "that I first learned the meaning of the word '*montage*,' a word that played such an important part in the development of our film-art." Pudovkin, *Film Technique*, 166. Imported from French, as it was being used by filmmakers and critics like Abel Gance and Jean Epstein, to describe the specific artistic possibilities of the technique, the term was very quickly and widely adopted, used in the most general sense to designate editing in an explicitly artistic mode. Balázs, for instance, in *The Spirit of Film*, published in German in 1930, describes editing as "a process for which it is significantly a French term, 'montage,' that has come to be accepted in the technical vocabulary of film." Balázs, *Early Film Theory*, 122. Yuri Tsivian considers the emergence of "montage theory," distinguishing between the terms "cutting," "editing," as more practical applications, and "montage," which is, as he says, "the practice of editing in the mirror of theory," noting that "the term itself came from the French word *montage* via Russian *montazh* and settled in English in or soon after the 1920s." See Yuri Tsivian, "Montage Theory I (Hollywood Continuity)," in *The Routledge Encyclopedia of Film Theory*, eds. Edward Branigan and Warren Buckland (London and New York: Routledge, 2015), 306.

47 Kuleshov, *Kuleshov on Film*, 50.

48 Sergei Eisenstein, "A Dialectic Approach to Film Form," *Film Form: Essays in Film Theory*, ed. and trans. Jay Leyda (New York: Harcourt Brace Jovanovich, 1949), 48.

49 Ibid. The essay has been published in a different translation, with subtle alterations in emphasis, so that the passage reads, "shot and montage are the basic elements of film. MONTAGE. Soviet film has established this as the nerve of film. To determine the essence of montage is to solve the problem of film as such." See Sergei Eisenstein, "The Dramaturgy of Film Form," in *Selected Works, Volume I: Writings, 1922-34*, ed. and trans. Richard Taylor (Bloomington and Indianapolis: Indiana University Press, 1988), 163; capitalization in original.

would," in that respect, he says, "be made a principle."[50] The true principles of montage are theoretical, conceptual, not material or physical. Still, it is precisely by first of all cutting strips of film into pieces, and simply sticking them together, that the aesthetic possibility of montage emerges. It is the nature of such possibility that must be explained, he argues, by a theory of editing as montage.[51]

Montage theory is perhaps the most significant account of editing, and continues to influence most subsequent accounts, which tend to trace the origins of film art to the basic procedures of cutting and joining, to the material and physical facts of editing, establishing a relation between technical simplicity and aesthetic significance, between the rudimentary arrangement of shots and the expressive power and potential of the cinema. "Editing," as Valerie Orpen says, in her recent book on the technique, "straddles the line between art and craft."[52] It can be easily dismissed as a mere technical necessity, though, given its simplicity. Editing was derided famously by Claude Chabrol, she notes, as the "washing up," coming at the end of the truly creative processes of scriptwriting and filming. For Orpen, though, editing is essential. It is, she argues, the "art of the expressive," the primary source of the communicative and representational power of the cinema, the importance of which can nevertheless be too easily overlooked. As a technique, she says, editing is often effectively *invisible*, functioning at what she calls a "subliminal" level. While less easy to see, less obvious, arguably, than the other basic techniques of filmmaking, Orpen argues that it is nevertheless the most important, explaining cinema itself, insisting that "a discussion of editing is not only possible but also essential to a greater understanding of the ways in which films make meaning."[53] This was appreciated most acutely, she says, by the Soviet montage filmmakers and theorists. "As Eisenstein, Pudovkin, and Kuleshov claimed," she observes in

50 Eisenstein, "A Dialectic Approach," 48.

51 It is on this point that the differences between montage theorists emerge, and it is in this essay that Eisenstein famously distinguishes his theory from others. It is clearer and more explicit in the Taylor translation. For Eisenstein, montage is "collision," as opposed to Pudovkin's idea of montage, to what Eisenstein calls the "'epic' principle," the "*unrolling* of an idea through single shots." He also dismisses the arguments of "the theoretically outmoded Lev Kuleshov," as Eisenstein describes him, "[who] regarded montage as a means for producing something by describing it, adding individual shots to another like building blocks." For Eisenstein, by contrast, "*montage is not an idea composed of successive shots but an idea that DERIVES from the collision between two shots that are independent of one another* (the 'dramatic' principle)." Eisenstein, "The Dramaturgy of Film Form," 163; emphases in original.

52 Valerie Orpen, *Film Editing: The Art of the Expressive* (London: Wallflower Press, 2003), 1.

53 Ibid., 119.

her conclusion, endorsing this basic claim, "editing is indeed the foundation of film art."[54]

Sam Rohdie has similarly argued that editing, or montage, is foundational. It is, he says, perhaps the most significant of what he describes as the "artifices of cinema," allowing it to *represent* reality, and not just mechanically record it, transcending mere imitation or mimesis.[55] Like Orpen, he credits the Soviet-era filmmakers with the initial and most important and enduring insights about editing. Kuleshov, he says, recognized montage as "the essence of cinema. … The only reality for Kuleshov was that created by the editing."[56] It was Eisenstein, though, Rohdie suggests, who was arguably the first to fully grasp the essential nature of editing, and to put his theoretical insights into practice in his filmmaking. "Fragments (shots)," in Eisenstein's films, Rohdie says, "are not joined to create a continuity nor do they refer to an interior unity of which the fragment is an essential part, but rather correspond to a need to demonstrate a relation or to organise a significance."[57] Rohdie insists that he is not offering a theory of editing, or montage, admitting that it is "difficult to generalise or extrapolate,"[58] that he will only, and more modestly, offer a series of short critical essays on the particular editing patterns in the films of some of the most significant directors in film history, who, he argues, recognized the elementary power of the technique.

There is a clear theoretical implication here, though, the assumption that editing is an essential, foundational part of the cinema, and that editing derives its value from some more fundamental material base, which provides some basic link to "the real." Rohdie offers a kind of generalization, derived from his particular examples, that suggests a theory of editing, that posits at the very least that its conceptual and aesthetic complexity derive from a technical simplicity. "Montage," he says, "simply is the joining together of different elements of a film in a variety of ways, between shots, within them, between sequences, within these."[59] From such a basic material fact comes the range of complex expressive effects characteristic of the most accomplished filmmakers, whose editing he analyzes. In the end, Rohdie's critical essays amount to an effective endorsement of the claim that editing is fundamental or foundational, and he seeks to reconcile Soviet montage theory with its supposed theoretical opposite, the cinematic realism of

54 Ibid.
55 Sam Rohdie, *Montage* (Manchester and New York: Manchester University Press, 2006), 137.
56 Ibid., 27.
57 Ibid., 33.
58 Ibid., 1.
59 Ibid.

André Bazin. He concludes with an explicit rejoinder to (or theoretical rehabilitation of) Bazin, famous, of course, like Kracauer, for his supposed resistance to a montage aesthetics. Rohdie argues, though, that the "ambiguity that Bazin valued as inherent to the real can only be found and revealed (as he recognized) by the artifices of the cinema"[60]—the most significant being editing, conceived of as montage. "All films, perhaps" Rohdie says, effectively offering a generalization, "and the best ones especially, are a response to the questions: what is the cinema? What is its essence? What can it do best?"[61] As he implies in his analyses, it is editing that is arguably the essential part of the cinema. The cinema is the art of editing. It is at its best in the form of montage.

These are, I would suggest, theoretical claims, but they are only implicit in Rohdie's argument, manifested, for instance, in his attempt to reconcile Eisenstein and Bazin. As I will try to show, such reconciliation is indeed possible, but only insofar as they both in fact reduce the cinema either to the fact of editing or to the fact of photography. Both, that is, are equally flawed, establishing a normative, deterministic relation between a specific technical or material aspect of the cinema, presumed to be the source of the cinema's identity as an art. Indeed, despite the apparent fundamental differences, they in fact share a more basic assumption about the nature of the photographic image, as the fixed, immutable raw material of the cinema, differing only in their response to it—as what the filmmaker should ideally try to overcome or to which the filmmaker should submit. For both Eisenstein and for Bazin, the question of the cinema is reducible to the question of editing, which is to be either fully exploited as a technique, in the face of a recalcitrant raw material, which must be forced into desired aesthetic form, or it is to be subordinated to the more aesthetically significant fact of such material recalcitrance. A common assumption unites these two supposedly contradictory theories.

Indeed, it is this assumption, and these two basic responses, that have structured most accounts of digital editing, understood as the specific site of the recent technological transformation of the cinema. The crisis of the cinema in the digital era is typically reduced to a crisis of editing, understood either as a moment of aesthetic opportunity and possibility, fulfilling the formalist promise of film art, or as the end of an aesthetic era, dashing the realist hopes for the cinema. From a formalist perspective, Lev Manovich argues that the effects of new digital technology on editing can be traced

60 Ibid., 137.
61 Ibid., 85.

back to Eisenstein and Soviet montage theory, noting the limitations of the theory, though, insisting that Eisenstein "ultimately focused on one dimension—time."[62] Digital editing, Manovich argues, by contrast, offers new possibilities, by allowing for the control of other dimensions. "By establishing a logic that controls the changes and the correlations of values on these dimensions, digital filmmakers can create what I call *spatial montage*."[63] Sean Cubitt adopts an essentially Eisensteinian principle, arguing that there is "an internal logic at work in the genesis of cutting as the basis of film form."[64] That logic, discovered originally by Georges Méliès, Cubitt says, with his stop-motion effect, achieved through rudimentary editing, was pursued in most theoretical and practical detail by Eisenstein and the Soviet montagists, leading ultimately, he argues, to the comprehensive formal and manipulative powers of digital montage.

There is, on the other hand, the realist response to digital editing. D. N. Rodowick argues that the new digital editing systems, what he calls digital compositing, or the "digital event," provides an unprecedented degree of control, fundamentally transforming the film as a work of art, which now, he says, "is a fully imaginative and intentional object."[65] Film had always allowed for some degree of artistic freedom or latitude, Rodowick admits, but it was necessarily constrained. "Though subject to all manner of manipulation and transformation," he says, film has its origins in what he describes as "analogical automatic causation," which "serve as limits to subjective decisions and acts—they lend an objective, or as Bazin would have it, inhuman quality to the production of image."[66] This is what has been overcome, though, he says, primarily with the advent of digital editing systems. Like Rodowick, Dudley Andrew has argued that the new digital editing technology threatens what had been unique about the cinema as an art, by overcoming all (or most) constraints on intentional, imaginative control. "Films," he says, in their original form, actually made out of *film*, "exhibit tension between the human (imagination, intention) and the recalcitrant chunks of recorded reality."[67] He champions the film theory of Bazin, what he describes as his "theory of 'impure cinema,'" which, Andrew says,

62 Manovich, *Language of New Media*, 156.

63 Ibid., 158.

64 Sean Cubitt, *The Cinema Effect* (Cambridge, MA: MIT Press, 2004), 43.

65 D. N. Rodowick, *The Virtual Life of Film* (Cambridge, MA: Harvard University Press, 2007), 169.

66 Ibid., 49.

67 Dudley Andrew, *What Cinema Is! Bazin's* Quest *and its Charge* (Malden, MA: Wiley–Blackwell, 2010), 30.

"negotiates between man and nature, the imagination and the real."[68] This has been replaced, though, he argues by a "pure" cinema, fully imaginative, fully intentional, limited by nothing other than artistic will, no longer the site of such metaphysical or ontological negotiation.

In either case, for the formalist and the realist, there is an acknowledgement that the full realization of electronic video editing, in the form of digital systems, computer programs, has comprehensively transformed the methods of film editing, altering the cinema, both argue, at a more basic, more profound level, transforming the essential raw material of filmmaking, that had necessarily constrained the work of editing. It was this basic, raw material, the filmstrip, with the photographic imagery printed on it, and the sound recordings affixed to it, that was thought to have established the possibility of editing as a technique, and which was the basis for the very idea of montage, as a theoretical concept. Lucy Fischer argues that the effects on editing of the incorporation of the "high-tech methods" of digital technology are at least comparable to but likely greater than the effects of any of the other major technological disruptions in the cinema. "While in the 1920s and 1930s the coming of sound created one kind of crisis in editing," she says, "the advent of new, high-tech methods precipitates another."[69] It is in this respect, Fischer argues, that the advent of digital editing, an aesthetics of digital montage, has created a crisis of unprecedented consequence for the cinema itself, altering its very "ontology," so that a new theory or "philosophy" of the cinema is needed.

* * *

While a new philosophy of the cinema may indeed be warranted, it is not, I will argue, because of any change to its ontology. It is just this sort of confusion on the matter that I will seek to address. To that end, I will be tracing the critical and theoretical discourse about the aesthetics of editing, in both its original and digital forms, considering the debates about the relation between editing as a specific technique, or set of techniques, and the art of filmmaking more generally. One of the most significant issues has been the very distinction between the cinema and the other arts, the assumption that the cinema was unique, a kind of aesthetic exception, and that the incorporation of digital technologies have effectively destroyed what had fundamentally distinguished cinema from other arts, specifically from

68 Ibid., 30–31.
69 Fischer, "Film Editing," 81.

the "visual" arts, like painting and drawing and sculpture, and from theater, but from the traditional arts more generally. Emphasis has been placed inevitably on that technique that seemed unique to the cinema, namely editing. No other art possessed a comparable ability to effect such sudden transformations, shifting from one shot to the next, through the simplest of means, with nothing more than a cut, the effects of which seemed at once so fundamental and yet so variable.

The same technique, reducible to the basic gesture of the cut, was the foundation for the coherent elaborations of spatial and temporal unity of the so-called "continuity style" of editing, as well as the formal and compositional disjunctions of avant-garde and experimental filmmaking, of "montage" cinema. Editing has been understood as the basis of the cinema in its "classical," realist form, but also the primary means for the creation of a formalist avant-garde alternative. In both cases, though, the aesthetic possibilities of the cinema were reducible to the specific capacities of editing. In whatever form the cinema took, the fact of editing was always understood as the most definitive technique, understood, though, in relation to a yet more basic assumption about the origins of cinematic representation, of filmic depiction or composition, which have for so long been thought of as essentially photographic. The photograph, the causally generated or "indexical" image, was what had determined the basic limits of editing, constraining editing as a technique, restricting the imaginative intentions of the filmmaker, while simultaneously imbuing it with aesthetic value. Indeed, its value seemed to be derived just from the limits on intention and imagination. While such imagery could be arranged in whatever order a filmmaker desired, through the very simple procedures of film editing, the image itself was thought to be immune to any more fundamental editing process. To edit was to arrange in a linear sequence or series an essentially stable element, the *shot*, or the basic photographic component, which itself could not, or not easily, be changed in any fundamental way.

This, of course, is what has changed. The concept of editing, as a craft, as a technique, and as an aesthetic idea, has been transformed by new digital technology for the alteration and manipulation of the image itself, the shot itself, with significant consequences for the cinema as an art. Yet this sense of profound change is based on assumptions that are largely unexamined, within the usual precincts of film theory, and which can be called into question. These are some of the most basic and long-established assumptions of the theory of film as an art, which need to be reconsidered, in order to more effectively account for the recent technological changes to the cinema, and specifically to editing. The first question that I will consider is the supposed fate of film editing, which is thought to be determined by particular technological changes, according to certain basic understandings about the relation between art and technology, some quite fixed assumptions in much film theory. I will then consider each of these in turn.

The first and most important of such assumptions is about *intention* in art, which, in most accounts of the cinema, of film art, is reduced to the fact of editing, understood as the manifestation of a properly limited form of artistic intention, characteristic of the cinema, necessarily constrained by the causally fixed nature of the photographic image. This is primarily what is supposed to have distinguished the cinema from the other arts, which are characterized by an unfettered intentionality, making those arts entirely subordinate to the artist's will and whims, the products of an all-encompassing artistic subjectivity. The cinema, by contrast, before the digital era, was thought to be characterized by a general objectivity, derived from an inherent, *non*-intentional basis. It necessarily had a more intimate and objective relation to "reality," which the artist could not easily overcome, except, of course, by simply abandoning the basic process of photography, in favor of some form of "animation," or a too flagrant use of special effects, and so on. In the most basic sense, the cinema's aesthetic value was supposed to have derived, to a very significant degree, from the intrinsic, material limitations on intention, overcome only, or most properly, and in importantly restricted fashion, through editing. Yet, as I will try to show, to argue that the cinema derives its uniqueness as an art from a basically non-intentional process is to misunderstand the very concept of art, of the artwork, or to misconstrue it, to fail to understand the necessary relation between the fact of intention and the formation of works of art. It is intention, as I will argue, that defines any object *as* a work of art, undertaken through the use of specific tools, a particular technical apparatus, with which any work of art is realized or formed.

The second complex of problems, then, that I think is raised by the changes to film editing, are those encompassed by the word *technology*, the question of the relations between the tools used by artists to create works of art, the specific forms of which are always changing, the choice of which to use or which to eschew always potentially open. Such choices manifest in the form of specific techniques, which seem, on the one hand, to be irreducibly specific and individual, but which also apparently derive from the material facts of any particular technology, providing the basic formal means for the fabrication of works of art, which can be recognized as particular instances of general kinds of art, but which are thought to possess a specific or even unique identity. The relation between technology and technique, though, and the question of artistic identity, as I will try to show, is much more complex than often thought.

How and whether any particular filmmaking apparatus determines or contributes to the realization of the identity of the cinema as an art opens on to broader issues. From the specific questions of intention and technology, then, we are led to the more general concept of *ontology*, the question, that is, of the very status of the work of art as a particular kind of object. As I will try to show, a theory of the work of art, of the ontology of the art object, must be general enough to encompass any particular form of artmaking, including the cinema, which cannot be understood as aesthetically distinct or unique, to be contrasted with the so-called "traditional" arts, at least not in the terms that have so often been proposed by so many film theorists. At the heart of a viable concept of the work of art is an understanding of the relations between the aesthetic and the technical, an understanding of the inevitable variety of means for the creation of works of art as, precisely, technological means, put to artistic or aesthetic ends. Whether such means are as simple as those used for traditional film editing, scissors and glue, a blade and some tape, or as complex as new digital editing computer workstations, aesthetically they are the same.

It is these three basic concepts, then—these conceptual groupings, or clusters of questions—around which this book is organized, namely *intention*, *technology*, and *ontology*, as aspects of a more general aesthetics. It is my contention that, in order to understand the kinds of technical changes to the cinema, and specifically the changes to the tools of film editing, given its particular significance as a technique in so much film theory, these conceptual categories require some considerable clarification. To that end, I will be surveying some of the main currents in film theory, and some of the most significant accounts of the effects of digital technology on the cinema, in light of the broader and more general work in philosophical aesthetics, which, as I will try to show, can be brought to bear on some of the enduring conceptual problems of film theory, specifically the problem of editing, offering the means for understanding and explaining the kinds of changes that have taken place recently in the cinema.

An aesthetics of digital montage, in this respect, is an account of the cinema as an art, understood, in the most basic sense, just as the site of the complex relations between intention and technology, between the aesthetic and the technical, as manifested in the undeniably significant effects that may be created through the specific and, arguably, unique technique of film editing. The ontology of a work of art must be understood to derive just from such relations, as the products of artistic effort with the means that are available, with tools and technology that are always changing, within conditions that are, inevitably, historically variable. In this respect, then, the cinema, as an art, as the art of filmmaking, cannot be reduced to any particular technology, to any specific material basis, or even any particular technique, but must, like any art, be understood as an artistic activity, as a kind of artistic making, undertaken with various and variable technical means, used in many different ways, to create what can be understood precisely as *works* of cinematic art, the products of such activity.

To say that this has not been appreciated by film theorists is not right. As a particular branch of the philosophical inquiry into the arts, as a part of the broader endeavor of the field of aesthetics, as I think it must be understood, film theory has in fact been especially sensitive to the particular problem of the relations between intention, technology, and ontology in art. Yet, and as I will try to argue, certain assumptions, long established within film theory, on these very issues, have become rather sclerotic, hardened into the form of accepted and often unquestioned truisms, rather than illuminating and explanatory concepts. Especially fixed are some of the most basic assumptions about editing, the art of film editing, which are being upset by the undeniably dramatic changes to the tools of the editor, and which invite, I think, a reassessment, a rethinking, which I will try to offer here, as I endeavor to describe those changes, and to propose an aesthetics of digital montage.

1. The Fate of Film Editing

Abstract
In this chapter, I offer an account of the relation between technological change to the tools of filmmaking, specifically the changes to the technical apparatus of film editing, and our theoretical understanding of the cinema as an art. While the cinema seems to have been changed by the transformation of the tools of film editing, it is rather the case, I argue, that certain basic assumptions of film theory have been shown to be untenable, limiting our ability to explain the technical change. A critical account of these theoretical assumptions will be the basis of an aesthetics of digital montage.

Keywords: Cinema, montage, theory, technology, digital, aesthetics

As a technical stage of filmmaking, editing has been utterly transformed, raising some new questions about the role and function of the technique, and reopening some old ones about its formal and aesthetic value. Film editing has been replaced by digital montage. It is, we may say, no longer simply "artisanal." It has become a much more elaborate, highly specialized, technologically intricate procedure, rivalling other techniques in complexity, but perhaps losing its distinctive qualities as a result. The manual work of cutting and splicing strips of film has been replaced by a computerized process, a digital procedure, performed with software on video monitors, virtually extending the possibilities of "editing," now broadly and expansively conceived, to practically every aspect of cinematic representation. Given the importance of editing, generally assumed to be the technical basis of film art, widely thought to be the aesthetic foundation of the cinema, a significant if not definitive technique of filmmaking, it is not surprising that a quite passionate debate has emerged about the effects of these changes on the cinema's identity *as* an art. Most accounts of a new "digital cinema," of the comprehensive transformation of the cinema as an art in the digital era, of the effects of the computer on traditional techniques of filmmaking,

Furstenau, M., *The Aesthetics of Digital Montage: Film Editing and Technological Change*. Taylor & Francis Group, 2024
DOI 10.5117/9789463722803_CH01

inevitably find these to be most acute in the changes to editing. While much else about the technical apparatus of filmmaking had changed, often quite considerably, over the course of the history of the cinema, before the digital era, the technology of editing remained more or less the same. Such stability, such technical simplicity, seemed appropriate for that technique that was supposed to be foundational. Indeed, the very simplicity of editing, its basic technical quality, was thought to be what made it foundational. Now, though, editing has changed, in a process that has been relatively sudden and thoroughly extensive, a comprehensive transformation of that part of the cinema that had seemed so resistant to change, and that had been so important to its artistic and aesthetic identity. Our basic understanding of the cinema itself seems to have changed, given the transformation of the technique of editing. This is the effect, though, I will argue, of certain assumptions about the cinema as an art, and about the aesthetically constitutive function of the technique of editing. A reconsideration of many of these assumptions is in order.

Editing has long been understood as fundamental to film art, or even as the very basis of the cinema *as* an art. Jean Mitry, for example, insisted that editing is an indispensable aspect of cinematic art. "The certainty of the rhythm and the balance of a film in effect depend on it," he says.[1] Mitry is concerned, though, to establish an important distinction. Strictly speaking, he says, editing is little more than "a meticulous and difficult task, work that has creative power, but is not creative in itself, with very rare exceptions, since a film is not constructed during editing."[2] He is making a principled distinction here, I think, between the material aspects of filmmaking, like the tasks of editing, and the more general fact of artistic creativity, between, that is, the technical and the aesthetic—a more general distinction that I will develop in this book. As important as editing is, Mitry argues, it must

1 Jean Mitry, *Esthétique et psychologie du cinéma I: Les Structures* (Paris: Éditions Universitaires, 1963), 354; my translation. "De lui dépendent en effet la sûreté du rythme et l'équilibre du film." The published English translation of the passage I quote is: "The whole rhythm and balance of the film depends on the editor's doing his job properly." See Jean Mitry, *The Aesthetics and Psychology of the Cinema*, trans. Christopher King (Bloomington and Indianapolis: Indiana University Press, 1997), 127. In my translation, I am trying to reproduce the slightly more impersonal tone of Mitry's original formulation, in which the pronoun "*lui*" is not referring to the person of "the editor," who is "doing his job properly," but to the work of editing itself, to the aesthetic effects of editing as a technique.

2 Mitry, *Esthétique et psychologie I*, 354; my translation. "Toutefois ce n'est rien de plus qu'une minutieuse et difficile mise au point, un travail qui a pouvoir de création mais n'est point créateur en ceci qu'à de très rares exceptions près on n'échafaude pas un film au montage." Cf. Mitry, *Aesthetics and Psychology*, 127.

be understood as only a part of a more complex aesthetic undertaking, that includes the original and intentional design of the shot itself. Still, it is editing that is often thought to be the specific technical source of the cinema's value as an art, the basis for cinematic *expression*, explained typically by analogy, by comparing editing, as the arrangement of shots, with language as the arrangement of words, with the linear assemblage of linguistic elements—an analogy that Mitry himself employs, but in interestingly critical fashion. "Where does this idea come from," Mitry asks, "that editing plays such a decisive role in film language?"[3] It is, he says, the result, "as always," of "a verbal confusion," a conceptual misunderstanding. "What is actually creative," he says, in contrast to the artisanal act of *editing*, "is *montage*, that is to say, what results from the association, arbitrary or not, of two images, A and B, which when related to each other produce an idea, an emotion, a feeling in the mind of the spectator, that does not exist in each of the images alone."[4] In French, Mitry establishes a distinction between "*montage conditionnel*," rendered in the English translation simply as "editing," and "*l'effet-montage*," translated as "montage." As developed originally by the Soviet-era filmmakers and theorists, which Mitry tries to clarify, the theory of editing as *montage* draws our attention to the difference between editing, "*montage conditionnel*," as a strictly technical aspect of filmmaking, and the *effect* of editing, "*l'effet-montage*," which is the intentional and creative use to which the technique is put.

It is a distinction that I will pursue, contrasting the technical components, the specific material operations and procedures of filmmaking, with the creative act, the aesthetic undertaking, the product of which is a cinematic work of art, an object, like any work of art, of aesthetic interest, which performs an aesthetic function. While implicit in Mitry's distinction, and in the Soviet montage theory that he endeavours to develop, I hope to make it more explicit, according to a more rigorous philosophical analysis of the aesthetics of the cinema, of the creative act of filmmaking, and of the film understood as a work of art. I follow the straightforward but significant definition provided by Gérard Genette, for whom "*a work of art is an*

3 Mitry, *Esthétique et psychologie I*, 354.; my translation. "D'où vient alors cette idée d'un montage conditionnel jouant une rôle déterminant dans le langage filmique? Simplement—comme toujours—d'une confusion verbale." Cf. Mitry, *Aesthetics and Psychology*, 127.

4 Mitry, *Esthétique et psychologie I*, 354; my translation; emphasis in original. "Ce qui est effectivement créateur, c'est *l'effet-montage*, c'est-à-dire ce qui résulte de l'association, arbitraire ou non, de deux images A et B qui, rapportées l'une à l'autre, déterminent dans la conscience qui les perçoit une idée, une émotion, un sentiment, étrangers à chacune d'elles." Cf. Mitry, *Aesthetics and Psychology*, 127.

intentional [intentionnel] *aesthetic object*, or, what comes to the same thing, *a work of art is an artifact* (or *human product*) *with an aesthetic function*."[5] It is the question of *intention* that is raised in most acute form by the technical transformation of the apparatus of film editing, revealing a persistent and deeply entrenched assumption in much film theory. The advent of digital editing, the emergence of what I am calling an aesthetics of digital montage, is often bemoaned, understood as the reduction or subordination of the autonomous quality of cinematic representation, captured causally by the operations of the camera, to an overweening intentionality, allowing the filmmaker to do whatever they want, without constraint or limitation, to the recorded images and sounds. The fixed photographic record has been replaced by the inherently malleable digital image as the basic raw material of filmmaking, changing what it means to edit.

"Editing a film today," as Dudley Andrew says, "invariably means transferring all audio-visual inputs into digital information that is then manipulated up to the point of the final cut."[6] While he admits that few complain about this undeniable technical advance, which makes editing so much easier in so many ways, it is, he argues, ultimately at odds with what he conceives of as a genuine cinematic aesthetic, which he hopes will persist in the digital era, but which he thinks is put at risk by the new technology. "There remain," he says, "intractable artisans, who, thinking of editing as a form of sculpture, need to touch celluloid and measure the length of shots in meters rather than in time-code."[7] The tactile quality of the material filmstrip, he argues, reminds the filmmaker of an innate quality of the cinema, which he worries will be lost in the digital era. There is an important tension, Andrew argues, between what he describes as the basic "raw material" of the cinema, and its manipulation through editing, or, as he says, "between the human (imagination, intention) and the recalcitrant chunks of recorded reality."[8] D. N. Rodowick has made similar claims, which he has developed in detailed and influential form. "The digital image," he says, "is more and more responsive to our imaginative intentions, and less and less anchored to the prior existence of things and people."[9] Digital editing, or what Rodowick calls "digital compositing," or "digital synthesis,"

5 Gérard Genette, *The Work of Art: Immanence and Transcendence*, trans. G. M. Goshgarian (Ithaca: Cornell University Press, 1997), 4.

6 Dudley Andrew, *What Cinema Is! Bazin's* Quest *and its Charge* (Malden, MA: Wiley–Blackwell, 2010), 51.

7 Ibid.

8 Ibid., 30.

9 D. N. Rodowick, *The Virtual Life of Film* (Cambridge, MA: Harvard University Press, 2007), 86.

he insists, "produces an image of what never occurred in reality; it is a fully imaginative and intentional object."[10]

It is my simple claim, though, that the cinematic "image," the work of cinematic art, has always been a "fully imaginative and intentional object." The technical changes to the cinema, the transformation specifically of the apparatus of editing, does not alter this fact, does not make the cinema any *more* intentional, any *more* imaginative, as Andrew and Rodowick argue, and as most film theorists assume. Rather, technical change must be understood according to a more general conception of the work of art, as Genette argues, precisely as intentional, as imaginative, as the result of the human use of any sort of tool or material, any kind of apparatus or medium, to create an aesthetic object, an object designed to perform an aesthetic function. As Genette says, "an intentional object can only be a human artifact, unless one puts forward a theological hypothesis (which I refrain from doing) about the divine intentions embodied in natural things."[11]

A great deal of film theory, if not "theological," assumes that an important part, if not the most important part, of a work of film art is necessarily embodied in it, in the photographic image specifically, as a kind of natural intentionality, often described as "indexicality," existing prior to, and even at odds with, the intentional efforts of the filmmaker, to which such efforts should properly be subordinated. It is this tension, this contrast, between the natural value of the shot, understood, as Andrew says, as an indexical "trace" of reality itself, and the film artist's efforts to "sculpt" this "raw material" into some deliberate form, but properly constrained by that reality, that seems threatened in the digital era. He develops the arguments of André Bazin, for whom, as Andrew says, "the finished film ... puts us in contact with reality," and who was, he says, "prepared to let the happenstance of nature more than the art of the editor determine the shape of the finished film."[12] Debates about the value of film art typically take this form, as a debate ultimately about the value of editing. It is editing, specifically, that has been understood as the site of intentional intervention in the cinema, as the means by which a filmmaker may impose an intentional design, but also where the "recalcitrant" quality of the shot, reproducing the recalcitrance of reality itself, will constrain such efforts. Such assumptions have determined much of the debate about the value of editing specifically, understood either as the means by which artistic intention is asserted, to fully "remold" the

10 Ibid., 169.

11 Genette, *Work of Art*, 4–5.

12 Andrew, *What Film Is!*, 41.

natural raw material, or as the site of constraint, a limited mode of intention, which allowed the natural value of that raw material to reveal itself. With the advent of a new digital apparatus for editing, there is an opportunity to revisit these debates, and to clarify the role of intention, the role of the imagination, in the creation of a cinematic work of art, to reconceive the art of film editing, to describe the aesthetics of digital montage.

***

Editing has long been understood as the source of artistic value in the cinema, the specific means for producing its most distinctive effect, the instant shift of perspective from shot to shot, the basis of a cinematic aesthetics. Editing is often understood as the technical foundation for the emergence of the cinema *as* an art, the technique that is unique to it, and the most powerful means of aesthetic transformation. Through the simple act of laying shots end to end, a qualitative change occurs, turning the raw recorded material of the cinema, the separate discrete elements, the single shots, susceptible themselves to no further modification to any significant degree, into a meaningful whole. The linear arrangement of shots, the basic fact of editing, has long invited comparisons with the ordering of the elements of language, suggesting a film grammar, or a cinematic syntax, from which meaning is generated. Mitry insists that, "to a certain extent editing can be said to be the art of expression and signification."[13] He pursues the linguistic analogy, arguing that, "montage (which is the basis for film language) is nothing more than another form of language morphology, where subject, verb, and complement have meaning only by virtue of their interrelationship."[14] Indeed, as he notes, in the "numerous

13 Mitry, *Aesthetics and Psychology*, 127.

14 Ibid. The notion of a film language, or a cinematic grammar, is one of the oldest concepts in film theory, based on the analogy between editing and syntax. It is a difficult to clarify such a concept, though, as Christian Metz famously demonstrated. "Seen from a certain angle," he said, "the cinema has all the appearances of what it is not. It is apparently a *kind of language* (*une sorte de langage*), but it was seen as something less, a specific *language system* (*une langue*). It allows, it even necessitates, a certain amount of cutting and montage; its organization, which is so manifestly syntagmatic, could only be derived, one believed, from some embedded paradigmatic category. … Film is too obviously a message not to assume that it is coded." Of course, as he demonstrates at considerable length in this famous essay, maintaining the terminology of Ferdinand de Saussure, there is no cinematic paradigm to be found, no cinematic "*langue*," but only, at best, a "*langage*," individual, idiosyncratic instances of filmic expression. Christian Metz, "The Cinema Language or Language System," in *Film Language: A Semiotics of the Cinema*, trans. Michael Taylor (New York: Oxford University Press, 1974), 67.

theories based on montage," editing is "held to be the key to the whole of film expression."[15]

It is in its expressive or signifying mode that editing is often called *montage*. Borrowed from French, derived from the verb *monter*, meaning "to assemble" or "to erect," it has often been used, most famously by the Soviet theorists, to designate the deliberate and intentional use of editing to expressive or more general aesthetic ends, with the implication that editing is the sole or at least primary means of expression in the cinema.[16] A central tenet of montage theory is that the artistic potential of editing derived from its technical simplicity, that the simple acts of cutting and reassembling were what made film art possible. "The joining of shots into a predetermined order," as Lev Kuleshov explained, writing in 1929, "from which a film is made is technically called *montage*."[17] From this he offers a basic formula. "Montage," he says, "is the organization of cinematic material."[18] For Sergei Eisenstein, editing, conceived of as montage, is the basis of a "film language," the "expressive means of cinema."[19] Such expressive potential, he argues, derives from a technical simplicity. The effects of montage are reducible to the simplest of means, to what he described as the cinema's "*defining* 'instrument of production': the scissors."[20] Such an elementary tool, such a basic implement, would remain the essential component of all subsequent editing devices, until the advent of digital editing. With only a pair of scissors, a blade, and a little glue, or some tape, shots could be cut out and

15 Mitry, *Aesthetics and Psychology*, 128.

16 For a good, brief history of the origin of the term "montage," see Yuri Tsivian, "Montage Theory II (Soviet Avant-Garde)," in *The Routledge Encyclopedia of Film Theory*, eds. Edward Branigan and Warren Buckland (London and New York: Routledge, 2015), 314–27.

17 Lev Kuleshov, *Kuleshov on Film: Writings of Lev Kuleshov*, trans. and ed. Ronald Levaco (Berkeley, Los Angeles: University of California Press, 1974), 47; emphasis in original.

18 Ibid., 48.

19 Sergei Eisenstein, "Film Language," in *Film Form: Essays in Film Theory*, ed. and trans. Jay Leyda (New York: Harcourt Brace Jovanovich, 1949), 112. Jean Mitry was explicitly trying to formalize and rationalize Soviet montage theory, acknowledging its importance, but endeavoring to give more substance to the analogy with language, given what he saw as the limitations of Eisenstein's theorizing especially, noting that "one would have preferred Eisenstein to have established the bases of a general aesthetic system instead of categorizing a very personal system of stylistics." Mitry, *Aesthetics and Psychology*, 128.

20 Sergei Eisenstein, "Béla Forgets the Scissors (1926)," in *Selected Works, Volume I*, 80. Eisenstein is arguing that other film theorists at the time, like Béla Balázs, whose account he is contesting, tend to 'forget" that the cinema is reducible to the simple means for editing, to such basic tools as a pair of scissors. Balázs, he says, mistakenly identifies the camera as the basic "instrument of production," rather than, and properly, scissors and editing, the shot as the source of meaning rather than the assembly of shots.

then rearranged, to realize the full aesthetic potential of the cinema.[21] The physical cutting of film makes possible the basic formal and conceptual principle of montage. "*The expressive effect of cinema*," as Eisenstein argues, "*is the result of juxtapositions*."[22] Such juxtapositions, when properly made, would allow the cinema to become an art on par with or greater than the other arts, an effect achieved so simply. The crucial cinematic act, Eisenstein argues, is indeed and quite literally the *cut*. "The conditions of cinema," he says, "create an 'image' [*obraz*] from the juxtaposition of these 'cuts' [*obrez*]."[23]

Debates about the relation between the technical process of editing and the aesthetic strategies of montage, and the more specific debates about the proper form of montage, may be set aside for the moment. If, though, we are to understand editing in the simplest sense, as the "organization of cinematic material," as the "juxtaposition" of shots that have been cut out and joined together, we need to explain what it means that the technical apparatus for such organization, the basic tools of editing, have been utterly transformed, with the introduction specifically of computer workstations and editing software, into "digital montage." This is, of course, only one part of the far more extensive changes to the cinema in the digital era, altering many of the most basic aspects of film production, but also, and very importantly, the structures of distribution and exhibition. Much has been said about the fate of the cinema as an art in the new technological context of "digitization." Indeed, the scholarly literature on the so-called "digital cinema" is now vast, and the claims about the effects of computer technology on film art have become very familiar—the most common being that the cinema, or the cinema as we have known it, has come to an end.[24] More recent changes,

21 Most accounts of editing begin with an emphasis on its technical simplicity, and the formal complexity that can nevertheless be produced. "The physical basis of film editing," as Noël Carrol has argued, "is the cut, the physical joining of two shots. We can easily account for this process with a little chemistry [and] some mechanics. … But editing involves more than chemistry and mechanics. It is a means of communication … a means of articulation whose practice enables filmmakers to convey stories, metaphors, and even theories to spectators." Carroll attempts (unsuccessfully as he later admits) to describe editing as "communication," but without relying on the "film/language analogy." I consider Carroll's argument in more detail later. Noël Carroll, "Toward a Theory of Film Editing," in *Theorizing the Moving Image* (Cambridge: Cambridge University Press, 1996), 403.

22 Eisenstein, "Béla Forgets the Scissors," 80; italics in original.

23 Ibid.; the original Russian nouns, the similarities between which Eisenstein is emphasizing, are included in the square brackets in the published translation.

24 For a useful survey of that literature, explicitly addressing just this claim, and offering some proposals for advancing what has become a rather moribund debate on the end or "death" of the cinema in the digital era, see Richard Grusin and Jocelyn Szczepaniuk-Gillece, eds., *Ends of Cinema* (Minneapolis: University of Minnesota Press, 2020).

including the advent of so-called "artificial intelligence," or AI, could have further and even greater effects.[25] Arguably the most significant change, though, has been to the technical apparatus of editing, and to the concept of montage. In order to explain what is happening to the cinema in an era of rapid and constant technical change, and what may happen to it still, we need to understand the transformation from film editing to digital montage. To do so means to return to some of the most basic debates in film theory, which are so often about the role and significance of editing, about the relative value of the technical fact of editing compared to the other components of the technological apparatus of filmmaking.

In these debates, a distinction is typically established, between editing, on the one hand, and the camera, or photography, on the other, between the subsequent arrangement of shots and their initial creation. While most montage theory tends to trace the source of artistic value in the cinema precisely to editing, others have argued that it is the technology of photographic recording that is the proper or true basis of the cinema as an art. Such a distinction is based on an assumption about the discrete nature of the two technical processes, contrasting editing and photography, montage as formal manipulation and the shot as basic raw material. This distinction has been confounded, though. In the digital era, editing is no longer just the simple arrangement of shots in a sequence, creating juxtapositions *between* shots, as the basis of expression and signification. The shot itself may now be "edited," as the compositional components *within* the image may be easily altered, modified, and rearranged. Digital montage provides the means for the control of every element of cinematic representation, enlarging and expanding the basic power of editing, with consequences for how we understand the expressive nature of film art, blurring if not destroying the most basic conceptual distinction at the heart of most film theory.

25 Christian Metz, writing in 1960, already imagined the potential effects of what we now call artificial intelligence on art. He cites Abraham Moles, who, he says, "anticipates a 'permutational art,' in which poetry, discarding the chaste mystery of inspiration, will openly reveal the portion of manipulation it has always contain, and will finally address itself to computers." Both structuralism and cybernetics, emerging at the time, promised a new aesthetic era, according to proselytizers like Moles. "The 'poet' would program the machine," Metz says, summarizing such a view, "giving it a certain number of elements and setting limitations; the machine would then explore all combinations, and the author would, at the end of the process, make his selection." It sounds like the often-proposed function of ChatGPT and other large language models of generative AI. Metz importantly distinguishes, though, between computer programming and what he calls "a distinct organizational act," which characterizes artistic expression as a human achievement. Metz, "The Cinema: Language or Language System," 37. See Abraham Moles, "Poésie expérimentale, poétique et art permutationnel," *Arguments*, nos. 27–28 (1962), 118-20.

The distinction has been blurred before, though, in theoretical accounts of editing. Conceived as montage, editing has in fact been thought of in more expansive terms. Eisenstein traced the fact of editing to the very effect of cinematic motion itself, arguing that "*all* cinema is montage cinema, for the simple reason that that the most fundamental cinematic phenomenon—the fact that the picture moves—is a montage phenomenon."[26] Individual images, film frames, are connected, in what he insists is an elemental editing process, what he describes as "micro-montage," arguing that "montage pervades all 'levels' of film-making, beginning with the basic cinematic phenomenon, through 'montage proper' and up to the compositional totality of the film as a whole."[27] The theory of editing, of montage, in order to advance, he argues, needs to understand the extensive and comprehensive nature of the technique. "So far," he says, writing in the 1930s, in a draft manuscript for a book-length study of montage that he never completed, "we have not treated the question of the shot as a montage sequence of frames. Now we perceive even the still shot as a montage process, as the first link in a continuous chain of montage that extends throughout the entire work."[28] The advent of new digital technologies of editing has made it even more important to understand the relation between this specific technique and the "entire work," the extensive nature of montage, broadly conceived, in the work of cinematic art.

Still, even Eisenstein posited a basic distinction between editing, or montage, on the one hand, and the material to be edited, the shots that would be juxtaposed, on the other. "The shot," he said, "considered as material for the purpose of composition, is more resistant than granite. This resistance is specific to it [and] has largely determined the richness and variety of montage forms and styles—for montage becomes the mightiest means for a really important creative remolding of nature."[29] If, for Eisenstein, this made editing the primary technique, the means by which this granite-like material could be subjected to the artistic will, an intentional and artistic "remolding of nature," achieved by the re-ordering and juxtaposing of shots, for others, most notably for André Bazin, this meant that editing must be subordinated to the fact of the shot's immutability, to what he famously described as the "ontology of the photographic image," according to which

26 Sergei Eisenstein, "Laocoön," in *Selected Works, Volume II: Towards a Theory of Montage*, eds. Michael Glenny and Richard Taylor, trans. Michael Glenny (London: BFI Publishing, 1991), 109.

27 Ibid.

28 Ibid.

29 Sergei Eisenstein, "Through Theater to Cinema," in Eisenstein, *Film Form*, 5.

it resists any attempt to remold it, imagining the photograph just as a kind of mold itself, which cannot and should not be easily changed. It is from the image, from the shot, from its origins in the causal process of photography, Bazin argued, that the specific value of the cinema as an art is derived. Despite the differences between the theorists of editing, the various montage theorists with their different emphases, he says, "they share that trait in common which constitutes the very definition of montage, namely, the creation of a sense or meaning not proper to the images themselves but derived exclusively from their juxtaposition."[30] On this basis, Bazin mounted his well-known critique of montage theory. "Montage," he says, "as used by Kuleshov, Eisenstein, or Gance did not give us the event; it alluded to it."[31] A basic contrast is established, which continues to characterize much film theory in the digital era, between a photographic realism, subordinating artistic will and intention to the causal function of the camera, and a montagist formalism, emphasizing the manipulation of reality through the deliberate act of editing.

In both cases, though, a clear contrast is established between the fact of editing and the material to be edited. The simplicity of editing, reducible to a pair of scissors, to a physical process of cutting, depends on the conception of a comparably simple material to be divided and rearranged, a clear distinction between editing as formal act and the shot as raw material—between the more basic categories of intentionality and causality. It is often argued that digital technologies have destroyed the integrity of the shot, erasing the basic distinction between its blunt immutability, the causally fixed camera recording, the raw material of the cinema, and the deliberate, intentional act of editing, the basis of artistic expression and deliberate remolding. The cinema is thought to have become as malleable and subjective as any art, destroying what had been distinctive about it as an art. An aesthetics of digital montage, though, would begin with an understanding of the cinema as precisely an aesthetic phenomenon, as an *art*. All art, of course, necessarily has its origins in technical processes. These, though, will inevitably change. Moreover, the value of cinema as an art does not derive from any particular technology, any specific technical process, either camera recording or editing, nor from any basic contrast or distinction between these, but from a more general relation between technology and art, between the technical and the aesthetic. It is that relation that I will consider as I endeavor to describe an aesthetics of digital montage.

30 André Bazin, "The Evolution of the Language of Cinema," *What Is Cinema?*, trans. Hugh Gray (Berkeley: University of California Press, 1971), 25.
31 Ibid.

***

Again, it is a simple historical fact that the technology of film editing has changed. Explaining what this means aesthetically for the cinema is no simple task, though. Editing, as Noël Burch said, is "so complex a subject," so difficult to understand, that it has effectively defied theoretical explanation. Indeed, as he admitted, and as I noted in the introductory chapter, we "probably do not yet have the means with which to undertake a serious analysis of it."[32] In the years since, there have indeed been very few efforts to try to explain film editing, to advance a theory of the technique, few serious analyses. There is, in fact, a common tendency to desist from such analysis, to simply accept that it may well exceed our ability to explain it. There is as least one basic principle, though, on which most seem to agree. Editing, as I have said, is often thought to be unique to the cinema, the primary source of the cinema's identity as an art—but the effects of editing have been difficult to describe theoretically. The technical changes to editing have revealed the limitations of the most common accounts of editing, challenging some enduring theoretical assumptions, re-opening perhaps the most fundamental debate about the cinema, those that have been advanced, challenging what is perhaps the most basic distinction at the heart of much film theory. As Daniel Yacavone has noted, there has long been a fundamental "dichotomy" in film theory. There have been two basic views, as he explains, so that "(genuine) film art consists of working on (or with) *either* the physical, material world and its given perceptual appearance *or* the images, sounds and signs already abstracted from it (via the camera) and waiting to be creatively organized into meaningful and expressive forms on the editing table—or, today, ... a digital, nonlinear editing suite."[33] Such new technologies have had what seem like direct effects on our understanding of the cinema, altering the basic premises of the debate. As Yacavone says, "digital technologies and processes ... call this dichotomy into question."[34] Digital *editing* technologies, especially, as he suggests, are thought to have had the most significant and confounding consequences. It is no longer so clear what counts as the basic "raw material" of the cinema, or what the limits are on the filmmaker's creative manipulations of it. As Yacavone says,

32 Noël Burch, *Theory of Film Practice*, trans. Helen R. Lane (London: Secker and Warburg, 1973), 47.

33 Daniel Yacavone, *Film Worlds: A Philosophical Aesthetics of Cinema* (New York: Columbia University Press, 2015), 61; emphases in original.

34 Ibid.

the traditional dichotomy can be traced back to "the difference between the classic realist and formalist positions in film theory."[35] This difference can arguably be reduced to the question of the relative values of photography and editing, whether the cinema's identity as an art was derived from the creation of its raw material, causally, through the photographic process of image-making, or intentionally, in the linear arrangement of that material by the editor. While the original debates between these two schools of traditional film theory were thought to have been settled, or to have been overcome, by so-called contemporary film theory, they have returned with the advent of a digital cinema, and specifically in accounts of the new modes of digital editing.

What has been revealed is that the question of the source of the cinema's artistic value, the realist quality of the photographic image or the formalist arrangement of shots, had itself never been settled, that the question of which of the basic techniques of filmmaking was primary or definitive had never been resolved. To address the new technical fact of digital montage, then, inevitably means returning to the most basic distinction between photography and editing, and the structuring function of that distinction in film theory. Victor Perkins argued long ago that film theory was "the embodiment of twin mystiques, one of the *image* and the other of *montage*."[36] If theorists could be advocates for one or the other, finding the source of the cinema's artistic identity in either the photographic process of image creation or in the deliberate arrangements of those images, in either photography as causal registration or in editing as intentional manipulation and reconfiguration, the most recent technological transformations have confounded this basic distinction. Yet, and as Perkins argued, it had always been a false distinction of "mystiques." If we insist on such a distinction, he says, "isolating" editing from all the techniques for the formal composition of the cinematic image, from all of the things "which change *within* the separate shots as well as between them," then, as a result, "we shall understand none of the elements (and certainly not the editing) because each of them derives its value from its relationship from the other."[37]

To understand editing, that is, as he suggests, is to explain its function within an aesthetic economy of relations, as a part of a complex technical and aesthetic whole. Yet editing is typically isolated, distinguished as a

35 Ibid.

36 V. F. Perkins, *Film as Film: Understanding and Judging Movies* (Harmondsworth: Penguin, 1972), 17.

37 Ibid., 23.

technique, only to be contrasted with its putative opposite, a tendency made even more acute in accounts of digital montage. Such a distinction would seem to imply some theoretical clarity, but in fact editing has largely defied theoretical description, understood as an especially difficult technique to explain. Citing Burch, Valerie Orpen has argued that, unlike the other material aspects of filmmaking, unlike the techniques of cinematography, sound design, and *mise-en-scène*, which can be easily seen, heard, and described, the art of film editing is more difficult to analyze, less amenable to explanation, effectively *invisible*. "Editing," she says, in contrast to the more visible and tangible aspects such as lighting, sound, camera movement, and color, "is far more elusive. Editing can be equated with movement but unlike camera movement or movement within the frame, it is difficult to pin down, to freeze, to control."[38] As difficult as it is to separate and describe, though, Orpen argues that it is the primary means through which the cinema becomes "expressive," the effect of the arrangement of the basic elements of the shots into a meaningful linear ordering. While difficult to see and to evaluate as a technique, its effects, she argues, are achieved through these relatively simple means, by cutting strips of film into smaller pieces and splicing them back together. It is this basic material and technical simplicity, though, that seems to be the source of the theoretical obscurity, which Orpen in the end does not really clarify, nor does she consider the digital transformation of editing, which was more or less complete by the time she was writing.

While Orpen does not consider the theoretical consequences of the technical changes to editing, Sam Rohdie is reluctant to offer any theoretical generalizations about editing at all. He argues, though, that editing, or "montage," the artistic use of editing, is among the most basic of cinematic techniques, that it "simply is the joining together of different elements of film." He does not pursue the theoretical implications of this claim, however implying that editing may well exceed our ability to account theoretically for it. Indeed, he explicitly eschews theorizing, offering only "specific instances of montage, not the application of a general model to specific films."[39] He insists that what he presents is "not a work of theory that films might illustrate, but a critical appreciation,"[40] a description of notable instances of editing, of montage, in the works of several significant filmmakers over the course of film history.

38 Valerie Orpen, *Film Editing: The Art of the Expressive* (London: Wallflower Press, 2003), 3.
39 Sam Rohdie, *Montage* (Manchester and New York: Manchester University Press, 2006), 1
40 Ibid., 2.

While much of considerable interest is revealed about editing in Rohdie's "critical appreciation," much is also left unsaid about the aesthetic value and status of editing, which must be addressed if we are to understand what it means that so much about the technology of editing has changed. To explain these changes will require some considerable theoretical clarification, despite the longstanding and enduring sense that editing defies such efforts. Editing must be understood necessarily in relation to the other techniques of filmmaking, as part of an aesthetic whole, overcoming the dichotomy that has for so long characterized film theory, and that continues to do so.

***

Inevitably, this raises the question of origins. While often said to be the most significant filmmaking technique, unique to the cinema, perhaps even the source of the cinema's identity as an art, the history of its development is obscure, matching its theoretical obscurity. Jean Mitry quite memorably said once that the search for the origins of editing as a technique is like wanting to find out how butter was first cut.[41] As André Gaudreault and Laurent Le Forestier insist, though, quoting this passage from Mitry in their recent book on the history of editing, the considerable research into the early years of cinema that has been done subsequently means that "*it is now possible* [*il est désormais possible*] to 'find the origins of editing,' or in any case to better understand those origins or rather some of them."[42] They insist, moreover, that an understanding of its historical origins provides some insight into the very concept of editing, describing its history as "the passage from practice to idea," arguing that from its mechanical or practical beginnings in a relatively simple technical process filmmakers have derived the complex artistic technique of editing, according to a historical process that is explicable, and on the basis of theoretical principles that can be clearly articulated.[43] As Gaudreault and Le Forestier suggest, any account of editing must indeed understand it as a "passage" from practical efforts

41 Jean Mitry, "Entretien intempestif: L'histoire en question," *Cinématographe* 110 (May 1985): 76. This is my rough English paraphrasing of the original passage in French: "Vouloir définir l'origine du montage c'est comme de vouloir préciser l'origine du fil à couper le beurre."

42 André Gaudreault, and Laurent Le Forestier, *De l'assemblage au montage cinématographique: Instauration et standardisation d'une pratique* (Montréal: Les Presses de l'Université de Montréal, 2022), 23; my translation; emphasis in original. The quotation by Mitry is the epigraph to the introductory chapter of their book.

43 "It is," they say, "the history of this passage—from the *practice* of editing to the *idea* of editing—that is the subject of this book." Ibid., 44; my translation; emphases in original.

to conceptual form, in terms, that is, of the relation between an artistic technique and its technical origins, between, more generally, and as I hope to explain, the technical and the aesthetic. I mean this in the most general sense, though, and not to establish a relation between a specific aesthetic value and any particular technique or technology, any particular physical basis or material substance, but as the grounds for the establishment of a relation, between the technical and the aesthetic, understood as the basis upon which we may appreciate and judge a work of art precisely as a *work*, the result of actual labour and effort, with the use of particular and various tools and materials.

Editing is, without a doubt, a central if not in fact indispensable technique, the technical basis for many of the most significant aesthetic effects created by film artists. There is also no doubt that the technology of editing has changed, in dramatic and comprehensive ways. Most editing in the cinema is now done on a computer, with digital systems, using complex software, as part of a more general aesthetic transformation. Indeed, the changes to editing are where the shift from a so-called "analog" cinema to a new "digital" cinema are thought to be most acute, where the change is most precisely and significantly manifested. To properly understand editing in the digital era, though, to appreciate the aesthetics of digital montage, it is necessary to distinguish more clearly between the material fact of editing as a technology, on the one hand, and the idea or conceptual account of editing as a technique, on the other, and to see the importantly continuous part of that history, to see how, as an idea, editing as a technique is indeed intimately related to the technical means for realizing it.

Something has certainly changed about that relation, in strictly material terms, as Gaudreault and Le Forestier have noted. "Until the obsolescence of the filmic supports (*des supports argentiques*)," they say, "announced by the arrival of the digital at the end of the twentieth century, the fundamental gestures of editing consisted of *cutting* celluloid and *scraping* the gelatine to then *stick* the pieces of film together."[44] The substantially physical or material nature of the process, as emphasized by their choice of verbs in this passage (in French, the verbs they use are *couper, gratter, coller*), has been replaced by the subtler, but no less material and technical, acts of typing on a keyboard and clicking a mouse or trackpad, manipulating images in electronic form on a computer monitor. In order to appreciate the aesthetics of digital montage, it is necessary to grasp editing just as an idea, as a concept, as an artistic technique, arising from the relation between technical practice

44 Ibid., 45; emphases in original; my translation.

and aesthetic idea, to appreciate the passage from the *practice* of cutting to the *idea* of editing, and from film editing to digital montage, as not only a historical process but precisely as an aesthetic matter.

I mean "aesthetic," though, in a quite precise sense, as itself a kind of *relation*, that can be established with any kind of object, fashioned from any sort of material. I am guided here by the definition of a "work of art," in the most general sense, as proposed by Gérard Genette, who argues that it is, first of all, an "artifact, or human product, with aesthetic function," which is to say, as he explains, "an object produced with the intention of arousing an aesthetic relation."[45] In this respect, he explains, "it is not some kind of object that makes the attention aesthetic but the privileged attention to its aspect that makes any kind of object aesthetic."[46] For Genette, it is important to understand that a viable notion of the *ontological* status of a work of art depends upon clearly distinguishing between the physical quality of a work, in whatever form it may take—either, using the terms established by Nelson Goodman, and redeployed by Genette, *autographic* arts, "consisting," as Genette says, "in material objects, like those of painting, of sculpture, of artisan architecture," and *allographic* arts, "consisting of ideal objects, like those of literature, of music, and of planned architecture"[47]—and the aesthetic function such works may perform. Indeed, works of art, performing an aesthetic function, objects with which we may establish an aesthetic relation, are works just insofar as they "transcend the object, material or ideal, in which they seem to consist."[48] As Genette goes on to explain, works of art "transcend the object in which they *reside* (rather than they *consist* in it) and which I call (to account for this restriction) their 'immanent' object."[49]

45 Gérard Genette, "From Text to Work," in *Essays in Aesthetics*, trans. Dorrit Cohn (Lincoln: University of Nebraska Press, 2005), 23.

46 Ibid., 23–24.

47 Ibid., 20.

48 Ibid. See Nelson Goodman, *Languages of Art: An Approach to a Theory of Symbols* (Indianapolis: Hackett, 1976), especially 112–23. For a discussion of the distinction in relation to the broader question of ontology, see Stephen Davies, "Ontology of Art," in *The Oxford Handbook of Aesthetics*, ed. Jerrold Levinson (Oxford: Oxford University Press, 2003), 176–77.

49 Genette, "From Text to Work," 21; emphases in original. Genette's argument is presented in summary form in this essay. It is worked out in much more detail by Genette in *The Work of Art*, and the subsequent volume, in which he completes his argument, *The Aesthetic Relation*, trans. G. M. Goshgarian (Ithaca: Cornell University Press, 1999). I refer later, and in more detail, to these two works, in which Genette establishes a basic distinction between the technical, material facts of "immanence," and the intentional, aesthetic fact of "transcendence." A similar distinction is offered, in less theoretical detail, by Jed Perl. All artistic action, of course, is undertaken with particular materials, Perl argues, which provide both freedom and constraint. All artmaking, he says, is a deliberate reconfiguring of some material. "The reshaping, which turns experience

Genette is seeking to expand or complicate the distinction from Goodman. The distinction is not an entirely clear one, but in the simplest sense, and as Goodman argues, an *autographic* work typically exists as a singular instance, a particular painting or sculpture, for example, while *allographic* arts are those that can exist in multiple instances, such as a musical or theatrical performance or a printed poem or novel. Cinema, in this respect, seems best described as allographic, although D. N. Rodowick has argued that the cinema's status as one or the other of these is "ontologically" ambiguous. Citing Goodman, Rodowick argues that film is characterized by a "difficult hybridity." While it may, on the one hand, be obviously allographic, reproducible as multiple copies, the "performances" of which are manifested in the projection of the film, the cinema is also, and for Rodowick more importantly, autographic, the manifestation of a kind of original moment of physical inscription, consisting of singular photographic imagery. This ambiguity, this duality of cinema, that Rodowick posits, is the source of what he describes as the ontological "problems" that he argues characterize the cinema. "The digital image extends these problems," he then says, "in another direction. Computer-generated images are not autographic," he argues, "for two reasons: as 'synthetic' images they cannot be considered the physical act of the author's hand, nor do they result in an end product."[50]

What we see here, of course, is another instance of the basic dichotomy in film theory, presented, though, as an ontological problem, as a kind of inherent ambiguity, that is supposed to be the source of the cinema's unique status as an art, rather than as an effect of the theoretical discourse itself. It is this unique ambiguity, this aesthetic instability, prized by Rodowick and others, that is worryingly thought to have been dissolved in the digital cinema, specifically with the rise of digital editing. The cinema is now just like any other art, or, more specifically, like any other allographic art, having lost its autographic quality, the ostensibly more important part of the cinema as an art. Yet, as Genette argues, this distinction, if it is coherent at all, cannot be the basis of a normative account of any particular art.[51] Moreover, this risks reducing the question of art to strictly material matters,

into art," he says, "is both artisanal (a matter of mastering the tools of the trade, whether words, colors, shapes, sounds, or movements) and metaphysical (a never-ending competition between the rival claims of authority and freedom). The metaphysical is embedded in the material." See Jed Perl, *Authority and Freedom: A Defense of the Arts* (New York: Alfred A. Knopf, 2021), 6.

50 Rodowick, *The Virtual Life of Film*, 14–15.

51 Gregory Currie has also criticized Goodman's distinction between the autographic and the allographic, arguing that "the distinction fails to distinguish any kind of art from any other." See Gregory Currie, *An Ontology of Art* (London: Macmillan, 1989), 15.

strictly to questions of immanence, obscuring the more significant matter of the establishment of an aesthetic relation, through what Genette calls transcendence.[52] To argue, as Rodowick does, that the cinema has become purely allographic in the digital era, that it has lost its autographic quality, mischaracterizes cinematic art. The distinction is not so simple, nor is it an adequate basis for understanding the shift to a digital cinema, by characterizing it as a shift from the autographic to the allographic.

An aesthetics of digital montage, then, is to be elaborated according to an understanding of the work of film art, the cinematic work, as what necessarily transcends its immanent manifestation in any particular material object, in any specific physical form. At the very least, as Genette insists, "one cannot say, in the full sense, that a work *consists* exclusively and exhaustively in an object," that "the mode and manifestation of works is not restricted to 'consisting' in an object. They have at least one other mode of existence, which is to *transcend* this 'consistence,' either because they are 'embodied' in several objects, or because their reception can extend far beyond the presence of this/these objects, and, in a certain way, survive its (or their) disappearance."[53] This is the case, for example, with "lost works," which can persist in memory, or in description, if no longer available in their original physical form, as Genette notes. It is, of course, and more importantly, also what allows works to be transmitted in various forms, a "film" as a video Blu-ray disc, or as a television transmission, which are just as legitimate as a 35-mm projection in a movie theater. This can be extended to a more general historical principle, so that what had once been the most common material or physical vehicle for a particular kind of art can "disappear," replaced by another, but without implying the disappearance of the art itself.

In this respect, whether the cinema is "analog" or "digital," whether works of film art consist of photographic imagery or computer simulations, becomes less significant than the aesthetic relation that we establish with the object, which performs its aesthetic function as an instance of cinematic representation, made available to us in whatever material form. Of course, and as Genette admits, this is not to suggest that "works do not consist in anything at all, but rather to recall that their existence consists in an immanence *and* a transcendence. A considerable number of obscurities,

52 Genette is aware of the possibly misleading associations of the word "transcendence," and is careful to note that, as he says, "I do not mean to give this term here, any more than elsewhere, a 'spiritual' or even a philosophical (for example, Kantian) connotation. I use it in its etymological (Latin) sense; which is eminently secular: to transcend is to go beyond a limit, to overflow a closed space." Genette, *The Work of Art*, 11, n. 19.

53 Genette, *The Work of Art*, 10–11; emphases in original.

disagreements, and, occasionally, theoretical impasses," he adds, "seem to me to stem from a rather widespread failure to distinguish between these two modes of existence."[54] This is especially so, I contend, within the precincts of film theory, where such a failure to make just this distinction is acute, particularly in many recent accounts of the advent of a digital cinema, and of the transformation of the technical apparatus for film editing. Some considerable "obscurities" and "theoretical impasses" are obvious in accounts of the new modes of digital montage, which tend to reduce the fact of the cinema too simply to a particular material form, to emphasize the very real technical differences between cutting strips of film and manipulating images and sounds on a computer, which are presented, though, as the basis for something much more significant, as an *ontological* change to the cinema itself.

Like Genette, but in slightly different terms, Joseph Margolis has also considered the tendency to reduce works of art to their material instantiation, to the stuff of which they consist, especially in film theory, where, he argues, there is often a "confusion of the physical and the cultural."[55] Arguing for what he describes as a "cinematic humanism," Margolis suggests that film art offers an especially acute insight into the value of art more generally, which is that "every 'technology' is, if viable, infused with the *sittlich* values of the historical society that uses it; and, second, among genuinely complex technologies (like film), the possibilities of the first are only gradually and creatively constructed as we transform ourselves by the labor of mastering its evolving possibilities."[56] Indeed, Margolis notes, among the arts, the cinema possesses a particular power, largely as a result of the techniques of editing, insofar as it can present us with what he describes as a "montaged world." He argues that "the filmic liberty of the editing process is the artistic counterpart of the interpretive liberty we cannot help applying in the transient discontinuities of our experienced world. Film trains us expertly to assemble the narrative unities of our own world in editing the consensual (montaged) recollections of our society."[57] Film, he argues, "is the metonymic form of our continuing enculturation, made more transparent, more accessible, to ourselves than by any other art. ... We see the world (imaginatively) through film; its variety and ubiquity compel us to

54 Ibid., 11. Genette is effectively offering the distinction between immanence and transcendence as a more subtle distinction than that between the autographic and the allographic.

55 Joseph Margolis, "Mechanical Reproduction and Cinematic Humanism," in *What, After All, Is a Work of Art?* (University Park, PA: The Pennsylvania State University Press, 1999), 115.

56 Ibid., 116.

57 Ibid., 120.

view the gradual transformation of our own world as constructed, variable, manipulated and manipulable, plural, evolving, and, above all, matched by a coordinately constructed interpretive skill that film itself entrenches."[58]

I'm not sure that I follow Margolis to some of his more effusive conclusions about the conceptual efficacies of film art, nor do I think, as he seems to suggest, that the cinema is reducible to the fact of montage, which risks re-establishing the traditional "dichotomy," but I agree with him at least that the strictly "physical" qualities of the cinema are not the specific source of its "cultural" value, that the ontological status of the work of film art cannot be reduced to any particular material object, but is derived rather from our growing appreciation of the uses to which a film artist may put any aspect of the complex and always changing technologies of filmmaking.[59] Among those uses, made possible by the more general fact of cinematic technology, is editing, or montage, an important source of the cinema's aesthetic function, one of the most important means (but not the only one) by which a film may invite us to establish an aesthetic relation with it. "Film instructs us," Margolis says, "in the analogies that hold between its own montaged worlds and the constructed narratives of actual life."[60] Films do so in their form as what he calls "utterances," relatively complex representations fashioned through the varied and varying technological means offered by the cinema. In this respect, of course, film is no different than any other art. "All of the arts," as Margolis notes, importantly, "are the constructed utterances of an enabling age: we study ourselves in studying the arts and we are thereby continually alerted in the sensibilities with which we continue to do so. Hence," he adds, "there cannot be any fundamental disjunction between our grasp of film and our grasp of any older arts, and we cannot understand

58 Ibid.

59 Margolis' account of the cinema as an art of montage is based in his more general "ontology" of the work of art, of its "peculiarity," such that "works of art," as he says, "are the products of culturally informed labor and physical objects are not. So seen, works of art must possess properties that physical objects, *qua* physical objects, do not and cannot possess. ... Hence," he concludes, "works of art exist as fully as physical objects but their existence depends on the independent existence of some physical object. Works of art, then, are culturally emergent entities, tokens-of-a-type that exist embodied in physical objects." See Joseph Margolis, "The Ontological Peculiarity of Works of Art," in *Art and Philosophy* (Brighton: Harvester Press, 1980), 22, 24. It is just such "ontological peculiarity," distinguishing the work of art, that Genette also seeks to explain. In this respect, the "peculiarity" of the cinema, which Rodowick argues makes it ontologically unique, is in fact a quality characteristic of all art, a more general "ontological peculiarity" that is common to every instance of works of art, and not the basis for a principled distinction between the "analog" and "digital" cinema, or between the cinema and any other art.

60 Margolis, "Cinematic Humanism," 124.

ourselves or others if the arts are not legible. For ultimately, they are us."[61] The theoretical task, then, is to make the arts "legible," just as works that we create, works that are created through the actual work of artists, and that we appreciate *as* creations, understood as the result of the deliberate use of some technical apparatus, some material elements, and of some particular tools.

***

Part of any such theoretical appreciation, of course, of such an aesthetics, is to acknowledge and to understand how the material conditions under which such work is done can change, how the contexts within which works of art are created can be transformed, and with what sorts of consequences for both the creation of and for our experience of that art. The art of filmmaking has been especially susceptible to significant technical change, but the most recent changes are thought to have been the most important, with aesthetic consequences greater than any others. Among these, the changes to the technical apparatus for film editing have been the most extensive and arguably the most significant. The basic technology of editing has been altered in almost every aspect, with what seem like obvious and important theoretical consequences for our understanding of the cinema as an art. We need, as a result, an aesthetics of digital montage, an account of the current state of editing, to address what has been a radical transformation from a material process of cutting strips of film into smaller pieces and reassembling them, a physically simple act, to a virtual mode of electronic editing, on computers with complex and powerful software, that allows for an unprecedented degree of control and manipulation.

Editing had long been conceived of as a separate, distinguishable technique, as the linear ordering of what was effectively unalterable material—or at least very difficult to alter—the shots on the filmstrip, the recorded images and sounds, understood as what we could call *dense* and *maximal* units of composition, which were not susceptible to any further internal modification to any significant degree. While shots could be arranged and rearranged, placed in whatever desired order, the content of the shot was more or less fixed. Editing was defined by such material constraint. A more comprehensive form of digital montage has now emerged, though, a more expansive and powerful form of editing, that seems to apply to even the finest detail within the shot, that extends beyond the merely linear

61 Ibid.

ordering of shots. The changes are certainly technically significant, but the sense among most film theorists is that this has had a more profound effect, altering the cinema in a more fundamental, essential way, subordinating all aspects of cinematic composition to the fact of editing, or to a more expansive notion of *montage*.

Rodowick has made this claim in perhaps its most explicit and powerful form, arguing that, "as constituted through digital capture or synthesis, the image is always 'montage,' in the sense of a singular combination of discrete elements. Even an unaltered digital still," he insists, "is already a work of montage in this respect."[62] More generally, Rodowick argues, "the continuity of automatic analogical causation is broken," so that "the spatial unity of the image in time can no longer be assured or attested to by the digital image," with the effect, as he says, that "the powers of indexicality are weakened and decentered."[63] The consequences, that is, of the technical changes to editing, are profound, altering the basic nature of the cinema as an art by reconstituting what had been the irreducible unit of cinematic composition and expression. As Rodowick says,

> there may be 'takes' in digital cinema, but there are no shots. Or rather, what was previously considered a shot has now become a highly variable element open to interactive manipulation at the most discrete levels. In this respect, cinema has become more like language than image, with discrete and definable minimal units (pixels) open to transformation of value and syntactic recombination.[64]

While the changes that Rodowick describes have indeed occurred, with undeniable effects on the ways that films are made, the conceptual framework through which he analyzes such changes is, I think, inadequate and often misleading, confusing the necessary and contingent effects of technological change on artistic value, and mistaking the origin of such value in a particular material constitution rather than the intentional, artistic activities of filmmakers. One aspect of such activity is the diversity of technical choices, in any art, but certainly an art as technically complex as the cinema. Indeed, one of the most significant limitations of much contemporary film theory is an inability to readily describe the relations between the aesthetic and the technical, between the inevitably variable

62 Rodowick, *Virtual Life of Film*, 166.
63 Ibid.
64 Ibid.

nature of the tools of the film artist and the value of film as an art. Such limitations, some basic confusions in film theory about this relation, have been revealed by the kinds of responses to the changes to editing, which tend to ascribe to technological change a specific and necessary effect on the cinema's very status as an art, according to an untenable account of the relation between the technical material of artmaking and the value of a particular kind of artwork, or indeed of a particular kind of art.

Indeed, there is a general tendency to trace the specific aesthetic value of an art like the cinema to a particular technical apparatus, or material basis, to a specific technological configuration, changes to which will necessarily alter or transform the cinema, *as* an art. But this is to confuse aesthetic or artistic value with a material quality rather than as an effect of the intentional *use* of any kind of material. A renovated theory of film, a new philosophical or ontological account of the cinema, does indeed seem to be in order, as Rodowick suggests, in light of the changes to editing specifically. This is less the effect of the technical changes themselves, though, or of any "crisis" that they may have precipitated, of any challenge to the "realism" of the cinema, than of some significant shortcomings of film theory itself, which has not been able to account for such changes, except in terms of loss and disruption, of radical or "ontological" transformation. The "crisis" of film editing, I will insist, and in this respect, is a crisis less of technical change than of theoretical discourse—a discursive and conceptual crisis. The effect of the technical changes to editing has been to disrupt some of the most basic assumptions about the cinema as an art, revealing some significant conceptual limitations of film theory, but also suggesting some possible clarifications. The effects of new digital editing technology, that is, require some considerable theoretical reassessment, some reconsideration of the basic terms of film theory, which must become more aligned with certain basic principles of aesthetics, with a more general theory of art, in order to properly understand the cinema in the era of digital montage.

This is not, of course, to deny that the technical changes to the cinema in recent decades, and especially the changes to the technical apparatus of editing, have not been extensive and wide-ranging, with enormous consequences for how films are made. While I am primarily considering the discursive or theoretical effects of the changes to editing technology in the cinema, the most immediate effects were more practical, and undeniably significant, altering many of the most familiar working procedures of filmmakers and editors, which had become well-established, indeed deeply entrenched, after many decades of technical stability. The basic technology of film editing had hardly changed over the course of the history of the cinema, until the sudden and thoroughgoing transformation from a simple and essentially manual process to a wholly digital one. While there had been many technical changes to editing before, some minor and others more substantial, the basic process had remained essentially the same. Editing was first undertaken with little more than a pair of scissors, or a blade, and some glue, or a bit of cement. Soon enough, more complex apparatuses were invented, like the Moviola, the first stand-alone, integrated and dedicated mechanical editing device, an upright system, with reels and a viewing screen, which was introduced in 1924.[65] Later, new devices like the Steenbeck, the most popular of the so-called "flatbed" systems, would become standard in the 1960s and 1970s. These allowed for finer and more direct control, and the incorporation of, for instance, additional viewing screens, and more working space generally. The basic mechanical procedures of editing, though, reducible to the physical processes of cutting and splicing, snipping and gluing, remained in place.[66] It was only with the emergence of complex digital editing software systems, with the shift from the essentially *linear* process of spooling lengths of film though the reels of the mechanical editors, either a Moviola or a Steenbeck, to the *nonlinear* procedures of first electronic and then computerized editing, turning editing into the arrangement and manipulation of images and sounds in virtual

65 For more detail on the effects of the Moviola, see Paul Monticone, "Classical Hollywood, 1928–1946: Editing," in *Editing and Special/Visual Effects*, eds. Charlie Keil and Kristin Whissel (New Brunswick, NJ: Rutgers University Press, 2016), 205, n. 9.

66 The transition later to the flatbed systems is described in chapters in Keil and Whissel, by Dana Polan, "Postwar Hollywood, 1947–1967: Editing," 85–86, and Benjamin Wright, "The Auteur Renaissance, 1968–1980: Editing," 105–7. Barry Salt offers an important account of the new "editing machines" introduced in the 1960s, when, as he says, "there was a major transformation in the type of machine used for viewing the separate picture and soundtracks during the editing process, particularly in Europe. Up to this point in America," he says, "the Hollywood Moviola or similar machines had been used almost exclusively." The transition to the flatbed systems, partly as a result of the specific needs of television editing, as well as various other particular changes, like the picture-synchronizer or so-called "pic-sync," an alternative to the flatbed systems, and more specific tools like the new kinds of tape splicers, most notable the C.I.R "butt" splicer, change many of the procedures and working protocols in the editing room. "The innovations in editing equipment" during this time though, Salt says, "had no influence on trends in the way films were dissected. Although the flat plate editing machines made the handling of large quantities of lengthy sync shots easier, the trend towards the use of very long takes in Cinema Vérité and Direct Cinema had already started before the new equipment became available, and in feature film-making people had always managed to handle the editing of long take films with the older equipment when necessary." Salt is sensitive to the more complex interplay between technological innovation and stylistic change. See Barry Salt, *Film Style and Technology: History and Analysis* (London: Starword, 1982), 345.

form, as data on hard drives, that the basic process of editing changes in a fundamental way, for the first time in its history.

While the transformation of editing technology has quite radically altered the film industry, it has also upset many of the assumptions of film theory, about the source of the cinema's identity as an art, of its specific aesthetic value. One of the most basic assumptions has been about the significance of editing, as a particular source of that value. Indeed, editing has seemed aesthetically and formally fundamental, in its original and traditional technological configuration. Robert Kolker has argued that film theory is ultimately reducible to debates about editing, its material basis, noting that "controversy over the structure and importance of the shot and the cut forms the bedrock of film theory."[67] That "bedrock" has fractured, for both filmmakers and film theorist, as the very distinction between cut and shot has been dissolved by the new technologies of digital editing, re-opening some of the most basic questions of film form. Indeed, so much has changed so quickly about the cinema, about the basic technology of filmmaking, that theorists have had trouble keeping up. Through the 1980s and 1990s, the pace of technical change was rapid, and, as Laura Kipnis noted at the time, "film studies has been somewhat slow to come to terms with a changing apparatus, or to theorize the shifts in film language and grammar that technological change seems to have so rapidly brought about."[68]

A kind of stability has been established, though. The shift to digital editing is now comprehensive, with broad consensus among film theorists that the "language and grammar" of film has been thoroughly transformed. Yet the understanding of the relation between technical change and film form, between the aesthetic and the technical, is still beset by certain conceptual confusions, and by a general sense of misgiving, in contrast to the views of most filmmakers, who have had to contend directly with the fact of technical change, but who tend to have become relatively well accommodated to the mode of digital montage, now that it is thoroughly established. Indeed, the last remnants of film editing—the procedures and techniques for physically cutting and reassembling pieces of actual film—have, for the most part, all gone, as film itself has been almost completely replaced by digital video. As Sonja Schenk notes, in the recently revised edition of *The Digital Filmmaking Handbook*, the quality of digital video "has become so good,"

67 Robert Kolker, "The Film Text and Film Form," in *The Oxford Guide to Film Studies*, eds. John Hill and Pamela Church Gibson (Oxford: Oxford University Press, 1998), 15.
68 Laura Kipnis, "Film and Changing Technologies," in Hill and Church Gibson, *The Oxford Guide to Film Studies*, 597.

that most filmmakers opt for it now over 35mm film stock.[69] Digital video, she declares, "is better, cheaper, easier,"[70] adding that "it's not just image quality that's improved, computers are better, too. Faster data processing and digital connectivity means that you don't have to build and outfit a computer with specialized hardware upgrades to ensure that it can handle digital video editing."[71] Indeed, a high-end consumer-grade computer and flat-screen monitor is all a filmmaker needs, rather than the elaborate dedicated systems that had first been used, such as the Avid system, and other dedicated digital work-stations. This, she suggests, is where the effects of digital technology are most pronounced. "Editing gear," as she notes, "is where the most dramatic simplification has occurred."[72] It is effectively consolidated in a single computer, even a laptop—by now perhaps even a tablet computer, like an iPad. Given such simplification, it is rare and only getting rarer that a filmmaker would edit on film, even if, as is still sometimes the case, they originally shot on film.[73]

The digital transformation is complete, and the basic "controversy" of film theory, as Kolker would have it, the debates about the relation between the shot and the cut, seemed to require radical revision, as that relation has been digitally reconfigured. At the very least, the nature of that relation has to be reconsidered. For Kolker, the basic object of theoretical analysis is, or was, the "film text," as he calls it, which "is unified by certain established ways in which shots are made and edited together."[74] The analysis of film form, he says, "concentrates on the two basic building-blocks of film, the shot and the cut, and on the structure that comes into being when the film is assembled, the combination of shot and cut that is the finished film."[75] The relative value of these basic "building-blocks," was being brought into

69 Sonja Schenk, *The Digital Filmmaking Handbook*, 4th ed. (Boston: Course Technology, 2012), 3.

70 Ibid.

71 Ibid., 4.

72 Ibid., 241.

73 Acceptance among filmmakers of the technical transformation is not, of course, absolute. Statistics suggest, though, that most filmmakers have shifted to digital video, even though 35mm motion picture film remains available, and some notable filmmakers still use it. The numbers are small, however. In the United States, for instance, for the last decade or so, there have rarely been more than about thirty feature films shot on 35mm or other motion picture film stock. See Vadim Rizov, "The 29 Features Shot, in Whole or in Part, on 35mm in U.S. Release Year 2023," *Filmmaker* 32, no. 2 (Winter 2024): 120. Rizov has been tracking the numbers since 2012, offering an annual tally, and has concluded that, while film will still be used by some, the vast majority of films are now digital.

74 Kolker, "The Film Text and Film Form," 15.

75 Ibid.

question, though, as the fundamental means by which a film is assembled were being transformed. Film theorists had noticed the change early on, and started to consider its effects on our basic understanding of the cinema as an art. The change was so comprehensive, so apparently significant, marking a very clear shift with profound consequences. "The advent of digital non-linear editing," as Laura Kipnis wrote in 1998, "has pretty much finished off film editing."[76] With such a sudden and comprehensive technological transformation, an aesthetic transformation was thought necessarily to result. The changes to editing, moreover, were understood as part of, or the most basic and significant manifestation of, a more general transformation of the cinema, as digital technologies were introduced into the apparatus, altering what had for so long been thought of as the fundamental material basis of the cinema, and upsetting many of the most basic aesthetic assumptions about its status as an art. As Kipnis says, the changes, the transition from what had been an essentially *photographic* medium to a *digital* one, "are spawning complex theoretical questions about the ontological status of the filmic image itself."[77]

The specific questions about the new technologies, about how digital cinema functions, compared to the way that motion picture photography works, she says, "open onto an array of impossibly large questions about referentiality and indexicality, onto questions about mimesis and realism."[78] The altered status of the film image, the spawning of such questions, Kipnis implies, were the result primarily of the changes to the technology of editing, noting that "even films that are theatrically released on film are almost universally edited electronically (either on tape or on a non-linear system)."[79] The changes, though, had been long underway, as Kipnis notes, and other theorists had started to speculate about their effects on our understanding of the cinema, and the need for a "new philosophy" of film, a "new ontology," as Lucy Fischer had suggested, was felt acutely.[80] What form this might take, though, what kind of new film theory might be required, Fischer and others left open, suggesting only that the consequences would be immensely significant for any theory of the cinema, any account of film art. Still, it was not yet entirely clear what sorts of formal changes might be effected by the still relatively new technology of digital editing, what an aesthetics

76 Kipnis, "Film and Changing Technologies," 598.

77 Ibid., 596.

78 Ibid.

79 Ibid., 598.

80 Lucy Fischer, "Film Editing," in *A Companion to Film Theory*, eds. Toby Miller and Robert Stam (Malden, MA and Oxford: Blackwell, 1999), 81.

of digital montage might even be. Writing myself around the same time, with Martin Lefebvre, about the newly emerging technology of digital editing, we suggested that "it may be that it is simply too early to predict the impact—if any—of digital editing on films themselves."[81] There was some indication, though, that new formal possibilities were emerging. At the very least, and as we noted then, shots of longer and longer duration were being created, not by filming a single scene in one camera "take," but by digitally compositing several takes into what only appeared to be a single shot. This sort of composition seemed new, and seemed to require a new kind of explanation, which had not yet been advanced. "Digital montage has yet to find its theorists," we wrote. "One thing is clear, though: confronted with the montage-filled long takes of digital cinema … film scholars will soon be forced to review and rethink models of film style that have dominated the discipline for at least half a century."[82]

***

Among the most important and consequential assumptions about the cinema, that has to be rethought, that has to be reconsidered, is that editing is unique to it, that editing is what primarily distinguishes the cinema from other arts, and that a change to editing will imply a more basic and fundamental change to the very nature of film art, that the cinema's identity as an art is tied in some specific and essential way to the fact of editing. Editing has always been understood, though, in necessary relation to another competing source of aesthetic value in the cinema, namely the camera, or, more specifically, photographic recording. There have, as I have said, been two basic views on the matter. Either the cinema is the art of editing, the source of intentional artistic control in the cinema, the technique with which meaning is made in the cinema, giving shape to the amorphous and artistically inchoate recordings of the camera, which in essence are merely mechanical copies; or, the cinema is the art of the camera, deriving its primary and unique aesthetic value from the naturally realistic imagery that is created automatically, and spontaneously, by the causal process of photographic registration, which necessarily limits and determines the

81 Martin Lefebvre and Marc Furstenau, "Digital Editing and Montage: The Vanishing Celluloid and Beyond," *Cinémas* 13, nos. 1–2 (Fall 2002): 71. See also my later consideration of digital editing, in which I offer an initial alternative account of the "ontology" of the digital cinema, of an aesthetics of digital montage. Marc Furstenau, "Film Editing, Digital Montage, and the 'Ontology' of Cinema," *Cinémas* 28, nos. 2–3 (Spring 2018): 29–49.
82 Lefebvre and Furstenau, "Digital Editing and Montage," 90.

scope and extent of editing, which is the technique meant merely to discover meaning in the camera image, rather than create it.

The contrast here, between creation and discovery, between intention and causality, is of course at the heart of the basic debate of so-called "classical" film theory, between "formalism," on the one hand, and "realism," on the other, the two competing schools of thought, which proposed conflicting accounts of the source of aesthetic value in the cinema, according to whether its powers of formal manipulation were emphasized or its capacities for automatic, realistic, photographic recording. While this debate had been, or was thought to have been, replaced by subsequent theoretical movements, by very generally speaking various kinds of semiotic and ideological analyses, it has returned, as the central debate about the fate of the cinema in the era of digital editing.[83] The new technology of digital editing has, it is argued, either finally and fully released the formalist, creative potential of the cinema, overcoming the limitations that had been placed on it by the fixity of the camera recording, or it is what threatens the realist identity of the cinema, as the triumph of editing leads to the destruction of the inherent aesthetic value of the photographic image, thought to have been the basis of film art. This, though, is a false distinction, and is based on what I think are certain basic misconceptions about the cinema as an art, about the very concept of art, about the film as a work of art. Rather than restaging the traditional debates about "formalism" and "realism," as most accounts of digital cinema and digital editing have done, a more informative account of the cinema's artistic identity, and about the source of aesthetic value in an art like the cinema, could be articulated, through a consideration of the advent of new digital editing technology.

Most accounts of technological change in the cinema, and specifically of the changes to editing, tend to reproduce the traditional debates about

83 As Irving Singer has importantly argued, the basic debates of "classical" film theory were never truly resolved, and have persisted, often expressed in other terms, even in more "contemporary" forms of film theory. Rather than advocate for one or the other of the original theoretical formulation, though, much less dismiss them, Singer seeks to "harmonize" the views, to produce a more pluralistic account of cinematic art, the basic aspects of which guide my approach here. "To formulate a theory that can use elements of both realism and formalism," he says, "we must recognize that cinematic transformations arise from human and organic aptitudes that have a constitutive, not just causal, role in the making of movies. In ways that are often unique to itself, film satisfies needs, affective as well as cognitive, that are ingrained throughout our species. These needs are basically the same as those that motivate us in everyday life. But like our view of reality, they, too, undergo transformation by the film experience. They are processed by artificial constructions, in the manner of all artistic expressiveness, and are frequently made more manageable, sometimes more gratifiable, than they might have been otherwise." See Irving Singer, *Reality Transformed: Film as Meaning and Technique* (Cambridge, MA: MIT Press, 1998), 5.

realism and formalism, depending on certain untenable assumptions about the cinema in its original, "analog" form, against which the digital cinema is contrasted. Such accounts often confuse the basic relation between the technical and the aesthetic, reducing the concept of aesthetic value to particular physical or material bases or to specific technical affordances. The cinema is in fact a necessarily heterogenous, complex amalgam of various technologies and techniques, that constantly change, and that produce all kinds of unanticipated creative possibilities for filmmakers, while presenting them with as many artistic problems, and material, institutional, and economic obstacles. If the cinema is to be understood as an art, and the film as a work of art, it must be understood, as I am proposing, as the result of the *work* of the film artist.

The conditions in which an artist works, though, typically change, especially the material and technical conditions of artmaking, as new technologies are introduced, old ones replaced, and procedures and protocols disturbed. The history of the cinema, as of any art, can be described as a history of successive crises, specifically technical crises, precipitated by such changes, yet, as Jean-Marie Schaeffer has argued, these tend often to be less a genuine crisis for the arts than a crisis of the discourse on the arts. Such discourse tends, as he notes, to offer itself as a solution to a crisis, but typically without addressing the many unspoken assumptions or faulty premises of much theory of art. A theorist, he argues, should be concerned with the coherency of their theory, but as the means with which we may effectively describe and analyze art, as part of our more general effort to understand its importance to us. Yet it is often concerned with the fate of art, with the fates of the arts, which, it is often supposed, need only a proper theory to be saved or indeed to thrive, to be set on the track to their proper destiny. Instead, though, Schaeffer argues, theorists of art should be providing the descriptive and analytical tools for understanding art, in whatever direction it may take, whatever goals it may pursue, rather than seeking to guide or direct art according to any essential understanding of its material constitution. The discourse on art is often beset by crises, which tend to be concerned with the coherency of claims about art, especially when these are challenged or undermined by material and technical change.

As Schaffer argues, though, this should be carefully distinguished from the autonomous realm of production. "As for art itself," he says, "I am convinced that it will get along very well on its own."[84] I am certain, too, that the art of

84 Jean-Marie Schaeffer, *Art of the Modern Age: Philosophy of Art from Kant to Heidegger*, trans. Steven Rendall (Princeton: Princeton University Press, 2000), 4.

the cinema will get along very well on its own, but am nevertheless concerned to account for the kinds of technological changes that such an art regularly undergoes, without proposing to determine, though, in any essential way, what the true nature of that art is, or to judge any specific change as more or less a threat to that nature. An understanding and an appreciation of an art, of the art of film, and the art of film editing, can indeed be based, and is perhaps more soundly based, on an understanding of its technical origins. Such an understanding, though, can tell us something about particular aesthetic accomplishments in an art, but not anything at all about the art *itself*, understood in some necessarily essentialist terms. Yet, and as we shall see, and despite qualifications and demurrals, there is in much film theory still an effectively essentialist tendency, which emerges especially in accounts of the changes to the technical bases of the cinema, and which is especially pronounced in accounts of the new technologies of editing.

"Periods of intense technological change," as D. N. Rodowick has said, "are always extremely interesting for film theory."[85] It is during such periods of change, he argues, with essentialist implications, that we are obliged to return to what he says is film theory's "primary question: *What is cinema?* The emergent digital era poses this question in a new and interesting way," he argues, "because for the first time in the history of film theory the photographic process is challenged as the basis of cinematic representation."[86] Yet this is *not*, I think, and as I shall insist, the primary question of film theory. There is, of course, a lot to say about the real changes wrought by the introduction of digital technologies into the apparatus of filmmaking, many of which Rodowick describes so well and carefully. To do so, though, does not depend on positing a putative identity, an essential photographic identity or "ontology," as Rodowick does, understood to derive from a particular material base or technology, the destruction of which has then to be explained, the replacement of which is supposed to have occurred, as that technology changes, putting the fate of the cinema as an art in question.

Rodowick argues that it is, in fact, digital editing that is the specific source of the technological transformation of the cinema's identity, arguing that the new mode of editing, what he calls digital compositing, "produces something ontologically strange or curious. What appears to be photographic, and therefore causal, is simulated and therefore intentional."[87] A balance seems to have been upset in the cinema, he argues, which had supposedly been

85 Rodowick, *The Virtual Life of Film*, 9.
86 Ibid.
87 Ibid., 169–70.

established between editing, on the one hand, and the camera, on the other, between the formal manipulations of montage and the realistic power of the photographic image—which he reduces to the more basic, "ontological" categories of the intentional and the causal, corresponding to editing and photography respectively, and which he equates with the technical distinction between the digital and the analog. We can, though, I think, question whether such an equation can in fact be made, and whether that aesthetic balance is what had in fact defined the cinema as an art, or if there is a better, more informative way of thinking about the relations between techniques, and between the technical origins of an art and the aesthetic value we find in art.

***

There is much to say about digital technologies and the arts, and even about the emergence of a distinctly digital kind of art, made specifically through the affordances of modern digital computers. "There is," as Katherine Thomson-Jones has argued, "a great need for more philosophical work on the digital arts."[88] Fundamental to such work, though, would be an effort to clarify the distinction between the analog and the digital, about which, she says, there is a "small metaphysical debate."[89] On the whole, though, she suggests, "the artistic implications of this distinction remain unexplored."[90] There is, she says, an extensive amount of speculation, by scholars in a wide range of disciplines, "about the ways in which digital technology is changing our understanding of and engagement with the arts."[91] There are many claims, as she says, about the loss of the differences between the arts, about the confounding of "our understanding of the distinct creative affordances of different artistic media," and claims that "we have lost trust in images while at the same time allowing them to influence our thinking and ways of seeing to an unprecedented degree."[92] Such claims, though, as she says, are often very general, and tend to be advanced without much careful

88 Katherine Thomson-Jones, *Image in the Making: Digital Innovation and the Visual Arts* (Oxford: Oxford University Press, 2021), ix.

89 Ibid. She notes specifically, David Lewis, "Analog and Digital," *Noûs* 5, no. 3 (1971): 321–27; John Haugeland, "Analog and Analog," *Philosophical Topics* 12 (1981): 213–26; Matthew Katz, "Analog and Digital Representation," *Minds and Machines* 18 (2008): 403–8; and Corey Maley, "Analog and Digital, Continuous and Discrete," *Philosophical Studies* 155, no. 1 (2011): 117–31.

90 Thomson-Jones, *Image in the Making*, ix.

91 Ibid.

92 Ibid., ix–x.

argumentation, but rather as theoretical assumptions. "Given their level of abstraction and potentially deep importance," though, she says, "claims like this are precisely the kind that philosophers seek to test and clarify."[93]

That is what she undertakes to do, but considering the very broad category of digital art, with an emphasis on the truly novel forms that have emerged, rather than on the traditional arts that have been affected in various ways by digital technologies, but which nevertheless retain their original identity as a discrete art form. The cinema is such an instance, an art that emerged before the advent of digital technologies, but which has undeniably been affected by them. There is no doubt that digital technologies have changed many things about the cinema, about filmmaking as a practice, about the distribution of films and their exhibition, and about the experience of viewing and appreciating films. As an art, though, the cinema has retained its basic identity. The claim that this identity has changed, or been profoundly altered, or even that the cinema as an art has been replaced by another kind of cinema, a digital cinema, a new form, with a new identity, is often based on a basic misunderstanding of that identity, of the identity of art as a more general category. There is, I insist, nothing ontologically strange or curious about the new digital cinema, if that strangeness or curiosity is thought to be the result of the advent of a thoroughgoing intentionality, if, as Rodowick argues, the claim is that the cinema had once consisted of some nonintentional aesthetic components, derived from the causal process of recording, supposedly automatic and minimally resistant to human intervention, but now consists of material that is all of the same "ontological order," entirely intentional. The cinema, in both its analogical and digital forms, belongs to a single "ontological order," namely that of intentionality, as the necessary and irreducible product of intentional artistic activity, of the creation of works of art, *as* works, of the deliberate use to unpredictable ends of the various and changing technical affordances of filmmaking, and of the conventions and techniques of the cinema, the manifestation of cinematic artistry. The practices of filmmaking, and the conventions and techniques used by filmmakers, have developed since the cinema itself emerged in an earlier era of dramatic technological change and transformation, and continue to develop in the digital era, in ways that can and should be explained, as we endeavor to understand the importance that we attach to the cinema as an art, which I understand as the basic task of a theory or philosophy of the cinema.

Film theory is that philosophical branch of film studies that should, as Thomson-Jones suggests, engage in the testing and clarification of abstract

93 Ibid., x.

speculation. A considerable amount of film theory, though, especially in the digital era, remains in what we could call a primarily speculative mode, seeking to discover the art's identity or essence, a definitive undertaking, as opposed to a more concrete inquiry into the activity of filmmaking as an artistic undertaking that can take a wide variety of forms. Rather than trying to define the cinema, to ask what the cinema *is*, an inevitably essentialist endeavor, I am more interested in what filmmakers have done and could do with the various new affordances offered by digital technology, specifically new editing technology, inevitably in light of—in relation to and often at odds with—what has traditionally been done in the name of cinematic art. I am concerned to counter many of the most familiar claims about the digital cinema, specifically the claim that the cinema has lost its aesthetic distinctiveness, and the related claim that its links to "reality" have been severed, or that its unique realism, derived from its origins in photography, has been compromised, if not destroyed. In order to do so, I have narrowed my focus, considering the specific cinematic technique of editing, often thought to be unique to the cinema, a distinctively cinematic technique, changes to which have been understood as a change to the cinema itself. This means inevitably contending with the most prominent arguments about the digital cinema, insofar as many of the most familiar claims about the digital cinema can be traced back to the debates about editing, about the proper use of editing as a technique, about its function as a technique thought to exist in agonistic relation to what is often supposed to be the more primary or more basic technique of photographic recording. As that relation has been so thoroughly upset by the new affordances of digital technology, it is an auspicious moment to consider the relation between technology and technique, and to test and to try to clarify the philosophical assumptions that underpin so much of film theory, through a specific consideration of the transition from the art of film editing to the aesthetics of digital montage.

To do so is to enter quite contested terrain, the site of a contest of values, thought to derive from different and specific parts of the cinema's technical apparatus. The question of editing has always been fraught, typically contrasted with photographic recording. The history of film theory could be reduced, arguably, to a debate about the relative values of the two components of the cinema, to where one stood on the question of their respective functions and appropriate uses, to the two "mystiques," as Victor Perkins puts it, of the image and of montage. Either one or the other, it has so often been claimed, is the true source of the cinema as an art; either one or the other should be emphasized or moderated; either one or the other is where the essence of the cinema is to be found. "The mystique of the image," Perkins says, "grows out of the classification of film as a visual art."[94] As a result, at first, "the decorative and expressive use of pictorial space was given precedence over the dramatic use of real space."[95] Eventually, though, he says, the "mystique of the image entered into the orthodoxy as soon as the facilities of the camera were converted into the 'demands of the medium.'"[96] Especially in the theory of André Bazin, Perkins argues, the orthodoxy of the camera, of the photograph, became, as he says, a kind of "purism," a defense of "realism" as the "vocation of the cinema," supposed to be derived as an effect of photography. "Despite Bazin's careful qualifications and disclaimers," Perkins says, "realist theory becomes coherent only if we identify the cinema's 'essence' with a single aspect of the film—photographic reproduction."[97] The cost of such "coherence," though, as Perkins insists, is a misunderstanding of the complex and heterogenous nature of the cinema, which cannot in fact be reduced to any such "essence." Indeed, and as he says, it was just as clear to other theorists that "the visual resources of the cinema," the camera, the photographic image, but also all the techniques of *mise-en-scéne* and cinematography, "were not … thought sufficient to refute the charge that film-making was a mechanical process."[98] If the cinema was to be understood as more than mere recording, with whatever embellishments to the image, which could not in the end disguise its mechanical origins, then it would be necessary to show, as Perkins puts it, that "the process of film-making allowed as great a control over reality as any other medium. Orthodox theory found this in the mystique of montage."[99] The source of creativity in the cinema was editing. "The idea that editing 'resembles a creation,'" says Perkins, quoting Sergei Eisenstein, "dominated the development of orthodoxy."[100]

For Perkins, though, the cinema can be reduced neither to editing nor photography, neither to montage nor to the image, without distorting the very idea of film art. It can only be understood, and necessarily, as a composite whole of a range of techniques, no particular technique or material fact being given precedence or primary or essential significance. Perkins argues against an *isolating* tendency in much film theory:

94 Perkins, *Film as Film*, 19.
95 Ibid.
96 Ibid.
97 Ibid., 39.
98 Ibid., 19.
99 Ibid.
100 Ibid., 21.

> If we isolate cutting from the complex which includes the movement of the actors, the shape of the setting, the movement of the camera, and variations of light and shade—which change *within* the separate shots as well as between them—we shall understand none of the elements (and certainly not the editing) because each of them derives its value from its relationship with the others.[101]

As I endeavor to describe an aesthetics of digital montage, it is just these relationships that I will emphasize, so that editing, in both its original incarnation and in its digital form, as one element of the work of film art, cannot be understood except in relation to all the others. The relation of the formal arrangement of the elements *within* the shot, through the various other techniques of filmmaking, and through the process of "decoupage," the procedures for the preparation of all of the elements of which each shot will consist, anticipating the editing together of the shots, and of the relations established *between* the shots, the juxtapositions but also the correspondences, spatial and temporal, thematic and compositional, despite what is often claimed about digital editing, has always been the basis of cinematic representation, the complex procedures of cinematic depiction. Still, it is worth acknowledging that editing, as a technique, has indeed undergone a comprehensive technological transformation. If the power of the computer has been applied to the art of filmmaking, it is, arguably, the technique of editing that has been affected the most. Indeed, the power of the computer is manifested most obviously and acutely in the new digital editing systems—where filmmakers work directly on computers, where they can exploit in the most direct way the specific technical affordances offered by the computer. A digital camera, in the most basic sense, is still, ultimately, a camera, capturing light and registering it as an image. Sound is recorded digitally, but in the ordinary fashion, through a microphone. Digital projectors still do no more, essentially, than cast that light, that image, upon a screen. The recorded sound is reproduced through speakers. The basic elements of the technical apparatus of filmmaking remain recognizable. Except editing.

***

The recent changes to editing in the digital are relatively easy to describe—at least in technical terms—as I have suggested several times so far. It is worth

101 Ibid., 23.

considering the changes in some more detail, though, before proceeding with more strictly theoretical matters. Originally, editing was an almost entirely manual procedure. It could hardly have been simpler. The basic or raw material was worked on directly, by cutting actual strips of film with a pair of scissors or some other kind of blade and then splicing the pieces back together with glue or tape. Lengths of film were run back and forth, first simply by hand, or on spools attached to the edge of a table or bench, and then later through somewhat more elaborate and dedicated editing devices, such as the Moviola or the Steenbeck. The editor would find the precise spot to make the cut, then the splice, joining one piece of film to the other, completing the edit. Repeating this, a film was constructed out of these various pieces, conceived of as "shots," arranged in more or less complex sequential patterns according to various formal, narrative, dramatic, and expressive relations. This had remained basically the same for the first hundred years or so of film history, the stable material and technical foundation of editing as an artistic technique, which emerged very early in the history of the cinema. Now, though, images and sounds are stored on a hard-drive, as digital information, which can be retrieved through a random-access process, called up with a stroke on a keyboard or a click on a mouse or trackpad, to be displayed on a computer screen as discrete files that may then be easily edited together following the procedures of the graphical user interface. In such a form, almost every aspect of cinematic representation is indeed amenable to a much more expansive process of "editing," allowing for fine-grained, precise manipulation at the basically atomic level of the "pixel" or the "bit." Once so simple, perhaps the simplest part of the filmmaking apparatus, editing systems are now arguably the most complex and powerful technology available to filmmakers. The changes are so extensive that it has been called a revolution, turning editing into one of the most technically intricate parts of cinematic technology, upending most assumptions about its value as a technique, as distinguished from the other techniques, the other technical aspects of filmmaking.

While a great deal has been written about the digital cinema, the primary emphasis has been on the replacement of what has been understood as the traditional raw material of the cinema, the basic celluloid film stock, on which photographic images were imprinted, eventually with a soundtrack affixed to the edge, with digital video, an electronic mode of recording imagery and sound—the replacement of the "film image" with the "digital image," understood as fundamentally different phenomena. As Holly Willis insists, though, in her account of the new digital cinema, this is "only one half of the digital filmmaking revolution. ... The other half is computer technology and the advent of desktop editing and visual effects tools that allow filmmakers to edit, animate, and otherwise alter digital footage on their own computers."[102] This is arguably the more significant technical event at the heart of the emergence of a digital cinema. Digital video recording is still just *recording*. It is the technology of computer editing that allows for such recording to be altered and manipulated. In some important respects—although I will question some of the underlying assumptions of this claim—the digital image can indeed be said to be more amenable to manipulation, compared to what is often called "analog" imagery, photographic recording, altering or destroying what is commonly described as the "indexical" nature of the photograph. In order to fully exploit the inherent and expansive manipulability of the digital image, though, new tools were required, in the form of new digital editing systems.

The extent and apparent significance of the changes to editing were noticed early on by film critics and scholars, having, as Laura Kipnis so bluntly put it, "pretty much finished off film editing."[103] The original tools of editing, with which editors simply cut and spliced actual film, were replaced with computer arrays and software programs, for the manipulation of digital information, first with dedicated professional systems, introduced in the 1970s and 1980s, such as the CMX 600, Lucasfilm's EditDroid, and the popular and powerful Avid system, and later with personal computer programs like Adobe Premiere or Apple's Final Cut Pro.[104] With these and other systems, the technology of editing was utterly and comprehensively transformed. If digital video is understood as more amenable than film to fine-grained alteration and manipulation, *after* the moment of recording, during the phase of post-production, the means for doing so are provided by new digital editing systems. It is here that the specific consequences for film art are made manifest, where the changes to the cinema are thought to have occurred, altering what it means to make a film and what a film means as a work of art. As Willis insists, new editing software programs "alter modes of production and aesthetics as well."[105]

It is to the aesthetic consequences that I will be turning my attention, which are generally thought to be profound. If, as Kipnis says, digital

102 Holly Willis, *New Digital Cinema* (London: Wallflower Press, 2005), 7.
103 Kipnis, "Film and Changing Technologies," 598.
104 For a very good, detailed chronology of the advent of new digital editing technology, see Michael Rubin, *Nonlinear: A Field Guide to Digital Video and Film Editing*, 4th ed. (Gainesville, FL: Triad Publishing, 2000), from which I derive these details, and the summary of the changes that I provide in the introduction.
105 Willis, *New Digital Cinema*, 8.

technology has indeed "finished off film editing," it also seems to have finished off film as an art, she suggests, or as a certain kind of art, bringing into question, as she says, "the ontological status of the filmic image itself."[106] The new technologies, especially the new editing technology, seem to pose a direct threat to the cinema, to its traditional aesthetic identity. As Kipnis says, "while there is no argument that computer-based random access editing ... is incredibly convenient and allows editors increased flexibility and creativity, it is still an experience fraught with risk."[107] That risk, she argues, is to the essential identity of the cinema as an art, which had been derived from the traditional tools and procedures of filmmaking, but which are being thoroughly reconfigured, indeed replaced by an entirely new set of tools. "Traditional filmic processes," says Kipnis, "are disappearing, replaced by new forms of digital image manipulation."[108] The specific site of such "manipulation," of course, is digital editing, the new computer systems for digital montage, where the aesthetic consequences of the changes to the cinema are most apparent and acute, opening on to what Kipnis described as "impossibly large" philosophical and aesthetic questions.

The technological changes to the cinema in recent years, during the digital era, have indeed been so extensive, so dramatic, so comprehensive, affecting almost every aspect of filmmaking, not to mention distribution, exhibition, and reception, that it is not surprising that the change is thought to be so significant, indeed fundamentally "ontological." It can indeed seem as though we are contending with an entirely new kind of cinema. It is certainly the case that the changes to editing technology are the most significant when compared to the other parts of the cinematic apparatus, which, over the course of the history of the cinema, have been subject to almost constant and substantial technological transformation. Digital changes are only the latest. Editing, by contrast, has been remarkably stable as a technology. For most of the history of the cinema, editing remained more or less the same. While there were developments in the design of the cutting apparatus, with the introduction of various kinds of dedicated editing devices replacing simple hand-held blades or a pair of scissors, upright or flatbed editing

106 Kipnis, "Film and Changing Technologies," 596.

107 Ibid., 598.

108 Ibid., 595. It is worth noting that Kipnis is in fact quite skeptical in her essay of the more extravagant claims for the "revolutionary" effects on the cinema of new digital technologies. Against the kind of claim that the digital cinema has been subsumed by a technocratic logic, threatening a truly cinematic aesthetic, she says: "Shooting film can *hardly* be considered less technocratic than shooting video, and in fact is far *more* technically difficult." Ibid., 600; emphases in original.

stations rather than a simple table or bench, and while there have been many other minor modifications, the basic physical facts of editing, reducible to *cutting* and *splicing*, were unchanged. The British filmmaker and editor Karel Reisz, the author of one of the most famous handbooks for film editors, first published in 1953, emphasized the very basic nature of editing as a technical undertaking. "The mechanical operations performed in the cutting room," he said, "are relatively simple. Unlike the cameraman or sound recordist, the editor does not require a great deal of specialised technical knowledge in order to be able to use his instruments."[109] Indeed, long after Reisz made this observation, through subsequent decades when the cinema changed so much in so many other respects, both aesthetically and technologically, editing hardly changed at all. "Compared to such technology-intensive crafts as sound mixing and cinematography," as Charlie Keil and Kristin Whissel have noted, "editing has been relatively unaffected by major technological developments."[110] Paul Monticone offers a similar observation. "The basic technology necessary to edit film," he says, "remained remarkably stable from the silent era through to the advent of digital editing systems,"[111] noting that while specific technical challenges emerged, particularly with the advent of synchronized sound, these did not affect the basic technology of editing itself, but rather had indirect consequences, altering working procedures and protocols, as he says, "altering the place of editing in the overall production process, as well as the editor's relationship with other craft people."[112] In the end, though, and despite having to contend with the additional element of sound, and with the new technical positions required for the fact of sound, editing remained basically the same as a physical procedure, now with the addition of a soundtrack.

Understood in the simplest technical sense, then, very little about editing had changed in any significant way since it first emerged as a technique. Strips of film were literally *cut*, with some kind of blade, and then *spliced*, reassembled in some preferred order, by gluing or taping the pieces back

109 Karel Reisz, with Gavin Millar, *The Technique of Film Editing*, 2nd ed. (New York and London: Focal Press/Hastings House, 1968), 393. This is a later "enlarged" version of the handbook, with additional material by Millar, and the first of many such updates after its original publication in 1953. These later versions noted the development of various new film styles and movements, and the effects on editing technique, while still emphasizing the stability and simplicity of editing as a technology.

110 This is from their "Introduction," in Charlie Keil and Kristin Whissel, eds., *Editing and Special/Visual Effects* (New Brunswick, NJ: Rutgers University Press, 2016), 6.

111 Paul Monticone, "Classical Hollywood, 1928–1946: Editing," in Keil and Whissel, *Editing and Special/Visual Effects*, 52.

112 Ibid., 53.

together. There was really little more to it than that. The technical simplicity of editing was long understood as the specific source of its aesthetic value. A relatively straightforward relation was thought to have been established between the material facts of the cinema, its technological constitution, reducible to a camera and a strip of photographic film, and editing as a technique, as the primary means for the subsequent manipulation of that material. There is some considerable debate about the origins of editing as a technique, but it seems to have emerged quite early, in the very first years of the cinema itself, an obvious development given the material quality of the filmstrip and the basic technological constitution of the filmmaking apparatus. The relationship between editing as a technology and as a technique seems quite clear historically. Scott Higgins, using the term in an especially expansive sense, dates the origins of editing to within a year of the invention of the cinema itself. "Editing," Higgins says, "conceived as joining two pieces of separately shot film, probably emerged in late 1896 with the advent of a loop that eased tension and allowed projectionists to assemble up to twelve single-shot movies on a single reel."[113]

Stephen Bottomore has argued that the cinema was conceived almost immediately as the art of the assembly of distinct parts, perhaps as early as 1896 or 1897, certainly, he says, within the first five years of the cinema's existence, derived from the experiences of those handling the basic film material. As a conception, he argues, it informed the style of even the earliest filmmakers, even before they started actually cutting and splicing pieces of film. While the very first films were, indeed, single-shot film, with no explicit editing, Bottomore admits, apparently made by simply starting a camera and filming whatever happened to be in front of it, he insists that "more care was involved than this would imply."[114] As he says, "from the very first films there was often a kind of 'editing' implicit in the process of shooting."[115] What was implicit soon became explicit, in the technical procedures of the cut and the splice, once editing was being done with a blade and some cement, with scissors and glue. Still, it was some time before filmmakers began deliberately cutting and splicing, and the full artistic and aesthetic potential of editing, as a technique for the arrangement of shots as discrete elements, became apparent. What we can recognize as editing was

113 Scott Higgins, "The Silent Screen, 1895–1927: Editing," in Keil and Whissel, *Editing and Special/Visual Effects*, 22.
114 Stephen Bottomore, "Shots in the Dark: The Real Origins of Film Editing," in *Early Cinema: Space, Frame, Narrative*, ed. Thomas Elsaesser (London: BFI, 1990), 104.
115 Ibid.

certainly evident, though, by the 1910s, and fully established as a technique by the 1920s, understood as the temporal arrangement of already composed images, pieces of film being put in new temporal order. Yet it is important to see, as Bottomore and others argue, that editing as an idea existed before filmstrips were actually cut into pieces, or, we might say, the *practice* of editing in implicit form, the consideration given to the careful arrangement of elements within the shot, came before the actual arrangement of shots, cut out and glued back together, and the establishment of relations between them. In contrast to the kinds of claims about the digital cinema, and digital montage, there is, as Bottomore suggests, no easy distinction to be made between editing as the establishment of temporal relations of shots and as the organization of elements of the shot itself. The composition of the images and their subsequent arrangement are best conceived of as a singular formal and artistic gesture, part of a single aesthetic act, understood as the intentional use of a particular technical apparatus, of a specific material form. It is to such a consideration that I now turn.

2. Editing, Intention, and the Work of Film Art

Abstract
New digital technologies for film editing have prompted a variety of critical and theoretical responses. In this chapter I survey several of the most influential accounts of digital editing, which tend to posit a certain conception of artistic intention, arguing that the new technology of editing effectively redefines the identity of the cinema as an art. Against such claims, I offer an account of the concept of the work of art, which is, I argue, characterized by a thoroughgoing intentionality, which persists in the transition from the "analog" to the "digital" era, and which cannot be reduced to any specific physical material, technological affordance or artistic technique.

Keywords: Digital, technology, technique, aesthetics, medium, ontology

The kinds of broad claims made about the digital cinema—about what distinguishes it from the traditional cinema that it is supposed to have replaced, about the referential quality of digital representation, about the veracity or reliability of digital imagery, about its effects on the viewer, about, in the most general sense, its artistic significance and aesthetic value, or its "ontology"—are often based on more specific claims about the expansion of the power of computerized editing. That expanded power is the direct result, it is often argued, of the loss of what had for so long been the material basis of the cinema, determining the formal possibilities of filmmaking, but also its limitations, and therefore the aesthetic value of the cinema, the loss, that is, of the essential raw material of filmmaking. "The celluloid strip," as D. N. Rodowick says, "with its reassuring physical passage of visible images, the noisy and cumbersome cranking of the mechanical film projector or the Steenbeck editing table, the imposing bulk of the film canister are all disappearing one by one into a virtual space, along with the

Furstenau, M., *The Aesthetics of Digital Montage: Film Editing and Technological Change*. Taylor & Francis Group, 2024
DOI 10.5117/9789463722803_CH02

images they so beautifully recorded and presented."[1] With these gone, or going, he wonders whether the cinema as we have understood it remains intact. "What is left," he asks, "of the cinema as it is replaced, part by part, by digitization?"[2]

Well, what's left, as he goes on to argue, is *just editing*—expanded to mean unfettered intentional control. Freed of what had come to be seen as the material constraints on cinematic representation, that "celluloid" filmstrip, which would first be run through the camera, and ultimately through the projector, having passed, importantly, through the spools on the editing table, has been replaced by the "digital image," which, Rodowick says, is "more responsive to our imaginative intentions."[3] Such intentions are more capable of being realized through editing, or through its expanded and extended form as digital montage. Editing is no longer conceived of as a necessary but inevitably intermediary or transitional moment, between recording and projection, as a specific and limited process during which the raw material of the camera could be modified, to a necessarily limited degree, and before the finished product would be projected, but as an expansive power that extends beyond the simple arrangement of shots in a sequence, to the unlimited capacity to modify elements of the shot itself, to an overwhelming and all-encompassing logic of modification and manipulation. The advent of the new digital systems is, effectively, the *triumph* of editing, of "montage." Indeed, Rodowick argues that, "as constituted through digital capture or synthesis, the image is always 'montage,' in the sense of a singular combination of discrete elements. Even an unaltered digital still," he says, "is already a work of montage in this respect."[4] The basic unit of editing, Rodowick argues, the shot, paradoxically disappears in the era of digital montage. As he says, "there may be 'takes' in digital cinema, but there are no shots."[5]

It is the "shot," as it has been traditionally conceived, the specific and indivisible unit of combination in film editing, that seems to have been destroyed in the digital cinema, as the most consequential effect of the "digitization" of the cinema, manifested in the form of an overarching logic of editing, with the result that there is effectively no longer any kind of constraint on the filmmaker. It is such constraint that is understood to have been the definitive aesthetic feature of the cinema, in its original form, the

1 D. N. Rodowick, *The Virtual Life of Film* (Cambridge, MA: Harvard University Press, 2007), 8.
2 Ibid.
3 Ibid., 86.
4 Ibid., 166.
5 Ibid.

basic material of which was photographic. Like Rodowick, Dudley Andrew separates the filmmaking process into three distinct, separate stages, what he describes as "the three phases, or moments, of transformation by which any film comes into being: recording, composing, and screening."[6] Each of these distinct and discrete phases, he says, "can be associated with its chief dedicated apparatus: the camera, the movieola [*sic*] (editing bench), and the projector."[7] The crisis in film theory, he says, "can be blamed on the modification, updating, or mothballing of each of these pieces of equipment, with twenty-first century digital technology, supplanting what is effectively nineteenth century machinery."[8] This transition to "the digital," he says, "is thought to perfect whatever operations its analog or manual predecessors were designed to perform; the digital enhances, expands, and alters these operations, achieving ultimate control."[9] He asks, too, like Rodowick, "what remains of the phenomenon of the cinema after the sweeping changes of the past two decades."[10]

Of all the changes, Andrew suggests, those with the greatest consequences for the cinema are to the technical apparatus of editing, expanded in its power, capable now of "ultimate control," overcoming what had been the strict material limitations imposed by the shot, that component of the cinema that traditionally resisted manipulation given its material constitution. "Films exhibit a tension," he says, "between the human (imagination, intention) and the recalcitrant chunks of recorded reality."[11] These are the "raw material," only partially susceptible to the imagination and intentional alteration. The editor, Andrew says, "operates with chunks of recorded material," arguing that "the basic unit of the film was unquestionably 'the shot' cut from the volume of the world, not the 'diaphanous image' created by the cameraman."[12] This is what has changed, with the most consequences,

6 Dudley Andrew, *What Cinema Is! Bazin's* Quest *and its Charge* (Malden, MA: Wiley–Blackwell, 2010), xxv.

7 Ibid., xxvi. Andrew has altered the spelling of the brand name of the first dedicated film editing device, the Moviola, invented by Ivan Serrurier, and made available commercially in 1924. For a brief history of the device, and the early years of film editing, see Paul Monticone, "Classical Hollywood, 1928–1946: Editing," in *Editing and Special/Visual Effects*, eds. Charlie Keil and Kristin Whissel (New Brunswick, NJ: Rutgers University Press, 2016), 51–67.

8 Andrew, *What Cinema Is!*, xxvi.

9 Ibid.

10 Ibid.

11 Ibid., 30.

12 Ibid., 36. Andrew is describing the arguments of Roger Leenhardt, the French critic, journalist, and filmmaker, whose writings in the 1930s, especially in the journal *Esprit*, were, as he says, such an influence on the thinking of André Bazin.

in the digital era. It is with editing that the expanded control of digital technologies is most acutely manifested, insofar as the "shot," the basic product of the camera, to be presented in sequences on the screen by the projector, has become amenable to complete manipulation and modification. Editing now, Andrew says, "invariably means transferring all audio-visual inputs into digital information," which can be too easily manipulated.[13] There are those, he says, naming Sean Cubitt specifically, who think that this is the fulfillment of the very *telos* of the cinema, understood precisely as the art of editing, of the cut, the original significance of which is manifested now in a material and technological prominence. "Editing stands as the medium's primordial 'special effect,'" says Andrew, summarizing Cubitt's view, "unchanged, Cubitt believes, even if now performed on AVIDs, rather than benches with synchronizers and rewinds. But," asks Andrew, "might not the new materials with which the editor works alter the very conception of that work?"

For Andrew, and for Rodowick, it does. The digital cinema is characterized by an increased, indeed comprehensive, *intentionality*, offering a greater scope for the imagination, which, for both, had been properly constrained in the cinema, in favor of what they effectively describe as the countervailing aesthetic value of the photographic image, the source of which is in an autonomous causal process rather than deliberate artistic intention. For others, such intentionality, the imaginative creativity of the film artist, is what had always been innately possible in the cinema, but suppressed, limited by the fixed quality of photographic representation, simply awaiting its ultimate fulfilment in the digital cinema, understood as the positive historical development of film art, or at least of its inevitable "destiny." That destiny is manifested, in inchoate form, in the very fact of editing. If, for Rodowick and for Andrew, the "conception" of the cinema may be changed by new digital editing technology, by the expansion of the power of montage, for Cubitt, it seems, it is a kind of realization or fulfilment of the inherent logic of the cinema, which is reducible to the logic of editing. Indeed, as Cubitt argues, in his history of what he calls the "cinema effect," it is with editing that the cinema achieves the status of art. The cinema emerges, he argues, in the exemplary early films of the Lumière brothers, as single-shot, single-take scenes, which are created without editing, but which establish the basic unit of cinematic composition, the separate images arranged on the strip of film, each image distinct from the other, which Cubitt calls, retroactively and anachronistically, the "pixel," conceived

13 Ibid., 51.

of in an historically expansive sense. Separated into individual, discrete images, arranged sequentially on the filmstrip, the photographed image is understood as a basic "picture-element," so to speak, as a pixel. These are first combined, in the simplest and most direct sense, merely by being projected at a certain rate, to create what Cubitt describes as the "first and most special of all effects" of the cinema, namely, the "depiction of motion."[14] In this basic fact, Cubitt finds the origins of digital cinema, in the necessary breaking down of the elements of an action, with the film camera recording, divided into separate images, discrete units, that could then be recombined, through projection, to create the impression of motion. "Movement," Cubitt says, in this respect, "starts in non-identity, the unstable zero pixel at its origin."[15] What's then necessary, though, he says, or what must necessarily follow, is editing, and with it the more complex unit of cinematic composition, arguing that it is "the cut that identifies the shot by defining its difference from other shots."[16] It is the cut, then, and the concept of the shot that emerges from it, as he argues, "that inscribes identity into the cinema."[17] The cinema is originally primarily spatial, he says, the boundaries, limits and artistic possibilities of which were first fully exploited by Georges Méliès, who created complex "tableaux," lengthy single shots, but which he then starts arranging in sequence, with a rudimentary process of editing, introducing a temporal dimension, or more specifically, a fully spatio-temporal effect. As Cubitt says, "these layered and contiguous spaces create the temporality of hiding and revealing and thus also the possibility of cutting from place to place."[18] Cubitt summarizes the historical shift from the early, simple, single-shot films to a logic of editing. "The innocent but abstracted delight in sensation," he says, "in light moving in time, gives way to the analytic pleasures of recognizing objects and their movements in space. This new mode of analytic cinema I will call the cut."[19] This came about in large measure by luck and happenstance, he says, recounting the familiar (and, as he in fact admits, likely apocryphal) tale of Méliès' camera jamming, and then restarting, creating the effect, when the results were screened, of a jump cut, a sudden transition in time, as a car (or a hearse, in some versions of the tale) that had been passing seems suddenly to appear. However he stumbled upon it, the technique of editing emerged to create

14 Sean Cubitt, *The Cinema Effect* (Cambridge, MA: MIT Press, 2004), 14.

15 Ibid., 41.

16 Ibid.

17 Ibid.

18 Ibid., 43.

19 Ibid.

ever more elaborate relations of space and time. "But," Cubitt insists, and despite the contingent nature of the discovery, "there is also an internal logic at work in the genesis of cutting as the basis of film form."[20]

Cubitt's rather eccentric re-telling of the history of the cinema, the complexities of which I will avoid discussing, consists, as I say, of a retrospective application of the primary elements of the digital cinema, the origins of which he finds in the very first moments of the cinema, in its initial manifestations, at the end of the nineteenth century, the basic elements of which are, as he says, the "pixel" and the "cut," to which he adds the third and final element, borrowed, as he says from the "language of computer imaging, the "vector." The "cut," he argues, "brought the principle of representation," and it exists, as he puts it, "in a dialectical relation to the flux of pixels without which it has nothing to organize, but which to it is always only nothing."[21] Thus the third element. "Out of the dialectical construction of object and subject a third principle arises: communication."[22] This is, he says, in the language of Saussurean structuralism, or Metzian film theory, "the paradigmatic axis of film. In deference," he says, "to its digital destiny, I will refer to it as the *vector*."[23] Again, we need not be concerned with the details of Cubitt's argument, or with the often-enigmatic formulations that he offers. What is more important is that, for Cubitt, as he suggests here, the cinema's ultimate "destiny," its inherent *telos*, is indeed digital (for better or for worse), which is why, as he says, he uses the language and terminology of "computer imaging" to describe the origins of the cinema. This destiny, though, can be reduced to his claim that cutting, or editing, the logic of combination, is indeed the "basis of film form." In this respect, and while he distinguishes himself from those who bemoan the coming of a digital cinema, his account is effectively as essentialist.

Others have made similar claims, most notably and influentially, Lev Manovich, who offers a more optimistic version of Cubitt's argument. Cubitt is, in fact, dubious about the digital cinema, which he argues is the fulfilment of the ideological effect of the cinema itself, what he describes as "the problematic of the cinematic object," the emergence of the "cinematic sublime," its identity as "the commodity form of communication in our time."[24] The cinema, he says, at least in its original constitution, offered

20 Ibid.
21 Ibid., 70.
22 Ibid.
23 Ibid.; emphasis in original.
24 Ibid., 11.

the "pretence of a timeless and universal order of communication."[25] Such "pretence" is now gone. "The digital," says Cubitt, "corresponds so closely to the emergent loss of an ideological structure to social meaning because it no longer pretends to represent the world."[26] Still, there is, from Cubitt's critical perspective, the possibility of cinema's ideological redemption, even in its digital form—indeed, he seems to be suggesting, just insofar as the digital makes so manifest the operations of the "commodity form," that it may, as a result, and in some vaguely Hegelian sense, possibly be resisted. I set aside speculation on such matters. The point is that, for Cubitt, the cinema is ultimately reducible to editing, to montage, whereas for Rodowick and for Andrew, it is to the photographic recording to the image. In these otherwise often abstract and enigmatic accounts, the traditional distinctions of classical film theory reemerge, between photographic reproduction, on the one hand, and editing, on the other, each proposed as the primary source of the cinema's artistic and aesthetic value, establishing the two oppositional poles of debate about the cinema, what V. F. Perkins described as the "twin mystiques" of film theory, "one of the *image* and the other of *montage*."[27]

Like Cubitt, and in contrast to Rodowick and Andrew, whom we may describe as "imagists," Lev Manovich is a digital "montagist." He argues that the possibilities of the "new media" are derived from the fulfillment precisely of the possibilities of editing, in the new form of what he calls "spatial montage," the complete realization, in digital form, of the possibilities of editing, first explored by the "montage" filmmakers, specifically the early Soviet-era filmmakers, placing at the center of the cinema an originally marginal avant-garde practice. In what he describes as the "archeology of compositing" in the cinema, Manovich says that editing had been limited or restricted mainly to the linear arrangement of shots, to the "technique of temporal montage," but that there had always been an alternative technique, of "montage within the shot," a spatial technique which was always possible, but much more difficult, and therefore "used more rarely throughout film history."[28] By "spatial montage," or "montage within the shot," Manovich effectively means composite photography, superimposition, optical printing, and other methods of combining various elements, two separate photographs, or a photographic recording and, say, matte paintings. It is also

25 Ibid.

26 Ibid., 250. "Digital media," he says, on the same page, somewhat enigmatically, "do not refer. They communicate."

27 V. F. Perkins, *Film as Film: Understanding and Judging Movies* (London: Penguin, 1972), 17.

28 Lev Manovich, *The Language of New Media* (Cambridge, MA: MIT Press, 2001), 148.

manifested in non-cinematic modes such as photomontage and collage. Always difficult to create and easy to detect in the cinema, it was with video that it became much easier, and less visible, specifically with the technique he describes as "keying," an electronic method for "combining two different image sources."[29] This is fully realized with "digital compositing," understood as the "continuation of montage within a shot."[30] While superficially similar to the kind of compositing that came before it, in cinema and in analog video, digital compositing, he argues, "does represent a qualitatively new step in the history of visual simulation because it allows the creation of *moving* images of nonexistent worlds."[31]

The basic process is fully manifested in the digital cinema, as a comprehensive logic of "spatial montage," which had only ever been marginal but is now central. It is this marginality that had limited the creative potential of film art, he argues, now released and unfettered. Digital technologies, Manovich says, in this respect, create the possibility generally of a kind of fulfillment of the cinema's aesthetic potential. "Computer media," he says, "redefine the very identity of the cinema."[32] The cinema had, he argues, originally been defined (and limited) by "its reliance on lens-based recording of reality."[33] No stylistic embellishment could overcome this fact. "No matter how complex its stylistic innovations," Manovich argues, "the cinema has found its base in these deposits of reality, these samples obtained by a methodical and prosaic process."[34] He offers what has become a famous formula. "The cinema is the art of the index; it is an attempt to make art out of a footprint."[35] As such, the possibilities for alteration, for creative and artistic intention, and for the most effective experimental and avant-garde strategies, were necessarily limited. Now that the cinema has been "redefined," though, those limits have been removed, so that "film obtains a plasticity that was previously only possible in painting or animation."[36] The most basic limitation, Manovich suggests, was the restriction of editing to merely liner and temporal arrangement. The new digital cinema expands the scope and power of editing, with the salubrious effect of opening up the artistic potential of the cinema, in the form of a new and expanded mode

29 Ibid., 150.
30 Ibid., 152.
31 Ibid., 153.
32 Ibid., 293.
33 Ibid., 294.
34 Ibid.
35 Ibid., 295.
36 Ibid., 301.

of editing. "Spatial montage," Manovich says, "represents an alternative to traditional cinematic temporal montage, replacing its traditional sequential mode with a spatial one."[37] The cinema, and other "cultural forms," he says, "become precisely a code."[38] As such, they are now, he argues, fully amenable to artistic intention, able to do whatever we want of them. In this respect, as he says, hopefully, "while new media strengthens existing cultural forms and languages, including the language of cinema, it simultaneously opens them up for redefinition," which is, he says, in the concluding line of his book, "the most promising cultural effect of computerization—an opportunity to see the world and the human being anew, in ways that were not available to 'a man with a movie camera.'"[39]

It is in this final implicit reference to Dziga Vertov, the early Soviet filmmaker and theorist, and to montage theory, the theory of editing to which Vertov contributed, manifested most famously in the filmmaking and writing of Sergei Eisenstein, that Manovich emphasizes his basic claim, which is that, in its original "indexical" form, even as employed by such an extravagantly experimental avant-garde filmmaker as Vertov, and the radical Soviet montage filmmakers, their efforts were necessarily limited, constrained by the fact of the camera, by the "deposits of reality" that it captured, that could not, in the end, be easily manipulated or modified to artistic ends. While Andrew may align himself in opposition to Cubitt, and to Manovich, in fact he and Rodowick tend to agree with them on the most essential points. The cinema, based in the process of photography, was limited—properly or regrettably—by that process, insofar as the results of the camera recording were "chunks" or "deposits" of reality itself, inherently resistant to intentional or imaginative alteration or reconfiguration. Digital technology, it is usually argued, has changed all this, transforming the basic material of the cinema into a wholly plastic, easily manipulated form of coded "information," with which the film artist may now do what they wish, without any material constraint, specifically through the affordances of digital editing technology. This is either bemoaned as a kind of loss, by Rodowick and Andrew, as the elimination of the photographic image as a unique form of representation, with an aesthetic value of its own, distinct from any other form of representation, or it is celebrated (if often cautiously) as "opening up" the cultural possibilities and artistic potential of the cinema as Manovich puts it, or as the fulfillment of a cinematic *telos*, as Cubitt

37 Ibid., 322.
38 Ibid., 333.
39 Ibid.

argues, in the form specifically of the expanded procedures of editing, of digital montage.

There is, though, I contend, a basic confusion at the heart of this debate, on both sides, which imagine creativity, imagination, artistic intention itself, as either constrained or released, limited or encouraged, by some specific technical affordance. Such an assumption, as I will try to show, derives directly from some of the most familiar debates in film theory, specifically about the relative value of editing, on the one hand, thought to be the specific and determinate site of intention and creativity in the cinema, and photographic recording on the other, the source of a rival aesthetic value, contrasted with intention and imagination, supposedly derived from the world photographed, from "reality" itself. This debate has never really been settled in film theory. It was, we may say, most explicit during the so-called "classical" era of film theory, in the argument between the "formalists" and the "realists." It was thought to have been overcome as a debate, or at least left behind, by "contemporary" film theory, by the structuralist, semiotic, psychoanalytic theory that emerged in the 1960s. It has returned in the digital era, specifically, I want to show, as a debate about editing, about the effects of the expansion of the powers of editing, at the expense, again, either for better or for worse, of photographic recording. The digital cinema, as Rodowick says, the image as either digitally captured or synthesized, is already "montage," before there is even any actual editing, the triumph of the "formalist" power of the technique, at the expense, though, of the "realist" quality of the photograph.

The conceptual boundaries of the debate, I argue, are severe and limiting, preventing any adequate account of the so-called "digital cinema," except in the most prescriptive and normative terms as either the proper or improper unleashing of intention and imaginative creativity, which are conceived of reductively as the direct manifestations of technical affordance. It is a limited understanding of intention in art, which is, admittedly, a fraught issue. It has so often been dismissed, in the various forms of anti-intentionalist arguments, as a necessarily vain pursuit of an author's or artist's intention, understood, that is, as the psychological state that the artist must have been in when the work was created, which presumably must be rediscovered and recreated, somehow, in the mind of the viewer or the receiver, in order to truly understand the meaning of the work. Such a psychological or "actual" intentionalism, though, is not what I am advocating.[40] Rather, it is something

40 Versions of such actual intentionalism, as we can call it, are offered, for example, by E. D. Hirsch, *Validity in Interpretation* (New Haven, CT: Yale University Press, 1967), and by William Irwin, *Intentionalist Interpretation: A Philosophical Explanation and Defense* (Westport, CT:

much more basic, as proposed, in clearest and starkest terms, by Gérard Genette, as part of his account of the concept of the work of art, which he has defined, in, as he admits, a "hypothetical or provisional" way. A "*work of art*," he says, "*is an intentional* [intentionnel] *aesthetic object*, or what comes to the same thing, *a work of art is an artifact* (*or human product*) *with an aesthetic function*."[41] As he says elsewhere, reiterating his basic claim, a work of art can be understood in the most basic sense, "as an object produced with the intention of arousing an aesthetic relation."[42] The definition, which he admits is not original, is worth restating and emphasizing, he thinks, given that "there is no lack of theories of art ... that neglect this aesthetic criterion, which seem to them not, or no longer, essential to the function of art."[43] The new forms of art, conceptual art, various kinds of postmodern or "contemporary" art, seemed themselves so obviously and flagrantly to flout any notion of intention, beginning, notoriously, with the "readymades" of Marcel Duchamp, that Arthur Danto has argued that art has become "posthistorical," and in that respect has left aside such vestigial qualities as "intention."[44] That artists may disguise, or eschew, or even reject their own intentions, though, in whatever artistic search or endeavor they are engaged in, does not mean that the theory of art must, necessarily, follow suit, Genette insists. Even in such minimalist or conceptual form as a "readymade," or a blank canvas, or an ephemeral instance of "performance art," all works

Greenwood Press, 1999). Richard Wollheim has described criticism in intentionalist terms, as a kind of "retrieval," arguing that the "task of criticism is the reconstruction of the creative process, where the creative process must in turn be thought of as something not stopping short of, but terminating on, the work of art itself." Richard Wollheim, "Criticism as Retrieval," in *Art and Its Objects*, 2nd ed. (Cambridge: Cambridge University Press, 1980), 185. A recent, broader discussion of intention, that tries to overcome the limitations of such "retrieval" models, in favor of a more general account of creativity in the arts generally, can be found in Berys Gaut and Paisley Livingston, eds. *The Creation of Art: New Essays in Philosophical Aesthetics* (Cambridge: Cambridge University Press, 2003).

41 Gérard Genette, *The Work of Art: Immanence and Transcendence*, trans. G. M. Goshgarian (Ithaca: Cornell University Press, 1997), 4; emphases in original. The original French term, "*intentionnel*," is included by the translator.

42 Gérard Genette, "From Text to Work," in *Essays in Aesthetics*, trans. Dorrit Cohn (Lincoln: University of Nebraska Press, 2005), 23.

43 Ibid.

44 See, for instance, Arthur Danto's appreciative account of a kind of hermeneutic, of deep interpretations, which "scorn reference to authorial representation altogether." There is, he says, "no end to deep interpretations, ... and who knows what deep structures the future will reveal. The artist's intentions have nothing to do with these." Arthur Danto, "Deep Interpretation," in *The Philosophical Disenfranchisement of Art* (New York: Columbia University Press, 2005), 49, 66.

of art are, ultimately and logically, intentional *acts*. And his theory of the artwork, Genette says, is, in the most basic sense, "intended to take into account at the same time, as much as possible, the work as *object* and the work as *action*."[45]

I will try to describe the intentional aspect of the work of film art, understood as a work of art, as both object and action. It is a thoroughgoing and comprehensive aspect of the work of art, I will argue, which cannot be apportioned to a particular part of the technical apparatus, manifested in the act of editing, but suppressed or countered in the process of photography, as is so commonly argued in film theory. An account of digital editing must understand it as a part of the work of film art just as an "aesthetic object," an "artifact with an aesthetic function," as Genette says, no different in that most general respect from any other kind of art, incapable of being divided into intentional and nonintentional parts. Such an account of the film, though, of the cinematic work of art as a kind of divided aesthetic phenomenon or object, has typically been proposed in film theory, according to the more specific distinction between the photographic image, as the raw material of the cinema, on the one hand, and editing, as the means for the intentional arrangement of that material, on the other. An account of the aesthetics of digital montage will, in part, I insist, have to overcome such a distinction, to understand the cinema as the site of artistic action, and of the creation of aesthetic objects, through whatever means chosen by the film artist.

***

This debates about digital editing, which I have just surveyed, were anticipated by Lucy Fischer's claims about the effects of the new editing technology. In her brief history of editing, of the various technical developments, and of the different ways in which it has been conceived theoretically, Fischer considers the latest technological changes, which she says have fundamentally altered the idea of editing, possible requiring, as she says, a new conception of the cinema itself. No longer limited to the arrangement of shots in a linear sequence, to the establishment of relations *between* shots, editing has been expanded to allow for manipulation *within* the shot itself. Editing is no longer limited to the ordering of shots as large, effectively irreducible units in linear or *temporal* sequence but has been extended to the arrangement of smaller *spatial* elements that comprise the

45 Genette, "From Text to Work," 26.

shot. Technological change in the cinema, she claims, and this change in particular, may require a change to our theory of the cinema, given what seems to be a fundamental alteration if not deformation of the art by new systems of digital editing. As she puts it, "in the era of virtual reality, we find ourselves in a virtual cinema—one that questions the 'integrity' of the shot."[46]

In a shift, that is, from mainly technical considerations to what we could call epistemological speculation, she argues that that integrity had been the source of a kind of "trust" in cinematic imagery, but that now "that confidence erodes," as we feel we can no longer believe what we see, given the expansion of the powers of manipulation, the advent of a new kind of *internal* editing. If the cinema had originally consisted of photographic imagery, that is, which resisted the modification of the material within the shot, that resistance, the source of a certain plausibility or reliability to cinematic representation, has now been overcome by new digital editing technology. With the expansion and extension of the powers of editing, in the new forms of digital montage, our faith or trust in a "realistic" mode of cinematic representation is eroded, if not destroyed. Indeed, realism itself, construed here as an effect that produces "faith" or "trust," may no longer characterize the cinema. It is here, she argues, that the radical effects of new digital editing technology, and the theoretical implications, can be seen most clearly. For Fischer, as we have seen, this means that "a new philosophy or ontology of the medium may be in order."[47]

Fischer does not say what form that "new philosophy or ontology" would take, but since she offered her initial observations on the potential effects of new digital editing systems, an enormous literature on the more general phenomenon of the "digital cinema" has emerged, often explicitly philosophical, offering specific accounts of the differences between the supposedly distinct ontologies of the new digital cinema and of the original, traditional cinema it is thought to have replaced, several of the most notable being those I have just surveyed. On the whole, there is a general consensus that the transformation has been profound, with enormous consequences for the cinema as an art, and for "realism" as a concept, understood to derive as a style from the cinema's more intimate relation to "reality" compared to other modes of representation, or indeed, as we shall see, compared to the very concept of "representation," against which the cinema, in its

46 Lucy Fischer, "Film Editing," in *A Companion to Film Theory*, eds. Toby Miller and Robert Stam (Malden, MA and Oxford: Blackwell, 1999), 81.
47 Ibid.

original constitution, is contrasted. If the "ontology" of the cinema had once been *photographic*, comprised of what is typically described as "analog" representations, the now very familiar argument goes, it has become a *computational* medium, made up of "digital" images. Such images, in their new digital form, it is claimed, are especially susceptible to modification, a susceptibility that is exploited, for better or for worse, by the new systems of digital editing, often isolated as the specific mechanism or technique through which the transformation to a digital cinema is realized. As Rodowick has argued, new digital editing, or what he calls "digital compositing," allows for something entirely new in the cinema, overcoming what had been the necessary limitations on artistic intention, the constraints imposed by an inherently realistic mode of generation, the spontaneous, automatic process of photographic recording, creating, as he says, "something ontologically strange or curious." In the past, he says, we were, as viewers, typically able to distinguish between the supposedly different causal and intentional components of a cinematic representation, when we could see, for instance, the boundary lines or the "seams" between a photographed scene and the matte painting superimposed upon it through the process of optical printing. The representations of digital synthesis and compositing, he argues, by contrast, "appear 'seamless' or perceptually real because all the elements now belong to the same ontological order—that of symbolization and its openness to intentional acts."[48] While the representations of the digital cinema, fabricated primarily by means of digital editing software, may look "photographically 'real,'" these have, as he says, lost the "indexical or causal qualities" that the real photograph had possessed. "Our previous perceptual criteria for realism," he argues, "have now ceded fully to imagination, fantasy, and the counterfactual powers of possible worlds."[49] Photography has become simulation, he says, and as such it "yields to a new imaginary that is unconstrained by causal processes; creation from physical reality gives way to the composition of 'elastic' reality."[50]

Many of the most significant accounts of the digital cinema tend, like Rodowick's, to emphasize the transformation of editing, as a technique that had been importantly limited in the past, constrained by the material quality of the analog cinema, but which has become effectively unfettered, ushering in a new era of cinematic art. If, as many claim, the analog cinema was characterized by the inevitable limitations it placed on the artistic

48 Rodowick, *The Virtual Life of Film*, 170.

49 Ibid.

50 Ibid.

intentions of the film artist, the digital cinema is thought to be thoroughly amenable to manipulation and fabrication, entirely intentional and wholly imaginative. Editing was long thought to be the specific technique through which artistic intention was expressed in the cinema, transforming the cinema's raw recorded material, awaiting meaningful form, into elements of expression or communication, through the establishment of relations between shots in a linear, temporal sequence. There have, as we have seen, been two basic views on the matter: one that saw editing as the means by which that raw, essentially formless material was given significance and formal meaning, and that celebrated the formalist powers of editing as the primarily expressive technique of the cinema; the other that saw editing as a necessary, indeed even inevitable technique, but which should be moderated in its use in order to allow the innately realistic quality of the raw material to express itself. In both cases, though, editing was understood as necessarily limited in its capacities, as the means by which shots were simply sequentially arranged, while the shots themselves could not be altered, or only crudely and with difficulty.

With the advent of digital editing systems, this distinction seems to have broken down, so that editing extends, as Fischer claims, and as Rodowick argues, into the very frame of the shot, the elements of which are as easily arranged spatially as the whole shot had been temporally, as the power of editing has been extended to the manipulation and modification of the content of the shot beyond the mere ordering of shots. The limits on modification, and the constraints on editing, were thought to be a limitation on intention itself, which was contrasted with the nonintentional origins of the shot in the causal process of photographic recording. If a new philosophy, a new ontology of the cinema, is in order, as Fischer suggests, it has typically been offered in the form presented by Rodowick, as an advocate of the second of the two views described above. For Rodowick, the effect of digital technologies, and specifically of digital editing systems, has been the comprehensive *reduction* of cinematic art to the *intentional* forces of the creative imagination, the loss of what had seemed to elude or escape intention, which had imbued the cinema with what is often described as a quasi-natural, *causal* value, the result of the supposedly automatic process of photography. In light of what appears to be the expansion and extension of the powers of digital montage, Rodowick argues, in apparently paradoxical fashion, the aesthetic effect is the "*devaluation of filmic editing*."[51] The loss or reduction of the value of editing is the direct effect, he says, of an increased

51 Ibid., 170; emphasis in original.

intentional power. The digital cinema is characterized by a comprehensive intentionality, in contrast to the analog cinema, which placed significant and necessary *limits* on the imaginative intentions of the film artist, constrained by the causally generated photographic image.

It is, I think, a peculiar way of thinking about the cinema, or of art in general. A new philosophy of the cinema, an account of its ontology as an art form, is indeed in order, I admit, but it should be a philosophy that understands that ontology as necessarily and irreducibly intentional, in both the so-called "analog" and "digital" incarnations of the cinema. While it may well have been more difficult in the cinema of the past to alter the photographic image after the fact of recording—but by no means impossible—and while it is certainly very easy now, with the affordances of the complex computer software programs, with digital editing systems, to modify and manipulate virtually any aspect of the shot, to replace and alter any element of a cinematic representation, the cinema is, and has always been, like any art, fundamentally an intentional phenomenon. The new digital technology does not make the cinema any *more* intentional, even as it does provide new opportunities for the manifestation of artistic intention, for the artistic imagination. The claim that the digital cinema, specifically through the means provided by digital editing systems, is indeed more intentional, assumes that intention can be plotted on a kind of sliding scale, and that a distinction can be made between the arts, and between particular historical and technological instantiations of an art, as more or less intentional, as an essential aspect of the specific material basis of an art. An account of digital editing in the cinema, of an aesthetics of digital montage, must confront such claims, in order to assess the genuine effects of the new technology, and to avoid what I think are some of the basic misconceptions of much digital theory. To that end, I will briefly consider the matter of intentionality, in order to clarify what I think is a certain confusion about it in much film theory.

***

Intentionality is a discrete and irreducible property of human activity generally, including the making of art, the creation of fictions and of imaginary worlds, if it is defined, as for example John Searle proposes, as a part of the effort that people make to "get mere objects to represent."[52] Indeed,

52 John Searle, *Intentionality: An Essay in the Philosophy of Mind* (Cambridge: Cambridge University Press, 1983), viii.

and as an initial and preliminary way of describing it, Searle says that, "Intentionality is that property of many mental states and events by which they are directed at or about or of objects and states of affairs in the world."[53] As part of a more general theory of mind, Searle argues that "Intentionality" is the expression or the "representation" of mental states, the production of utterances, or of any other kind of representative expression, that will be understood as the manifestation of an effort to make the content of mind, of mental states, publicly accessible. "The notion of representation," in this respect, as Searle says, "is conveniently vague."[54]

On the basis of this vague concept, Searle argues, "we can say that Intentional states with a propositional content and a direction of fit represent their various conditions of satisfaction."[55] If I say, for instance, that it is raining, the content of that representation can be "satisfied" in a certain respect, as the representation of a belief, by looking out the window, to confirm the fact that it is raining, or even by simply noting the raindrops on the window pane. If there is any doubt, I can simply go outside, to experience the rain directly, as it falls upon me. There is, then, as Searle argues, a basic relation established between representation and reality. The relation is logical, though, rather than ontological—and can be realized in any material form. When considering the more basic question of how language, specifically, as a material form, is related to reality, he says that the answer is that "language relates to reality in virtue of the fact that speakers so relate it in the performance of linguistic acts."[56] It is in this respect, as he notes, that his notion of representation "differs from its use in contemporary artificial intelligence and cognitive psychology. For me," he says, "a representation is defined by its content and its mode, not by its

53 Ibid., 1. Searle capitalizes the word "Intentionality," to designate it as the subject of his analysis, and to distinguish between the specific act of "intending" to do something, which is, he argues, only one specific instance or one form of the more general phenomenon of "Intentionality," which he defines as "directedness," rather than the simpler fact of merely being disposed to do something. As I proceed, my own use of the words "intentional" and "intentionality" will not be capitalized, but I will be using them in roughly the sense meant by Searle. The "directedness" of intention is an aspect of the broader phenomenon of "indexicality," which Searle discusses, a rather reductive version of which is used often by film theorists. Indexicality, we can say, following Searle, is that aspect of objects we get to represent that direct us to what they represent.

54 Ibid., 11.

55 Ibid.

56 Ibid., 197. In this respect, and as the attentive reader may have guessed, Searle's account of Intentionality is derived from speech act theory, especially that of J. L. Austin, with whom he had earlier collaborated.

formal structure."[57] This, of course, allows for a potentially infinite range of forms of representation, which are such in terms of content and mode, that is, "intentional content" and a "psychological mode," as Searle says. This is what makes them representations, which is to say predicated on conditions of (potential) satisfaction, as the establishment of the relation between the content of our representations and the world or reality that that content is *about*—as Searle insists. "Since all representation—whether done by the mind, language, pictures or anything else—is always under certain aspects and not others, and conditions of satisfaction are represented under certain aspects."[58] For Searle, this relieves us of the need to try to answer more traditionally "ontological" questions, about what belief or reality really *are*, by focussing instead on the content and mode of the representations of beliefs as representations of reality, "under certain aspects," that are of importance to whomever is creating the representation. So, as Searle says, "the question of how Intentional states are realized in the ontology of the world is no more a relevant question for us to answer than it is relevant for us to answer the analogous question about how a certain linguistic act is realized."[59]

Here we can begin to see the importance of all of this for any theory of the digital cinema that emphasizes an ontological difference on the basis of a difference in how what we can call, for the moment, a "cinematic utterance," comparable to a linguistic utterance or act in this basically logical sense, is realized. As Searle says, making the link to the cinema explicit:

> A linguistic act can be realized in speaking or in writing, in French or in German, on a teletype or a loudspeaker or *a movie screen* or a newspaper. But such forms of realization do not matter to their logical properties. We

57 Ibid., 12. A similar error is often made by film theorists, who, like many cognitive scientists, and artificial intelligence theorists, define representation by formal structure, often conceived of as, or thought to derive from, the material, technological basis of the cinema. These debates have recently become even more pressing, as developments in AI technology have accelerated, with the advent of so-called large language models, or LLMs, which are supposed to be able to produce "representations" more like genuine linguistic (or depictive) acts. Searle's distinction, though, still applies, I think. The simply formal structure of the sentences generated by ChatGPT are still significantly separated from content and mode. They often, as we have seen, do not match reality. The debates in film theory, in this respect, are suddenly more broadly relevant. We can see a tendency among many film theorists, who advocate for an autonomous causal quality of cinematic representation, to make a similar logical error as many advocates of a possible general AI.

58 Ibid., 13.

59 Ibid., 15.

> would, with justification, regard someone who is obsessed by the question whether speech acts were identical with physical phenomena such as sound waves as having missed the point. The forms of realization of an Intentional state are just as irrelevant to its logical properties as the forms in which a speech act is realized are irrelevant to its logical properties.[60]

Linguistic acts, as a particular type of representation of "Intentional" states, are one of many (possibly infinite number) of forms that such representation can take, that all, though, have the same logical properties, as Searle says. "The logical properties of intentional states arise from their being representations, and the point is that they can, like linguistic entities, have logical properties in a way that stones and trees cannot have logical properties … because Intentional states, like linguistic entities and unlike stones and trees, are representations."[61] Rodowick, though, for example, in his efforts to describe what he says are the "ontological" differences between the "analog" and the "digital" cinema, effectively argues that the former is indeed more like stones and trees, a kind of natural phenomenon, the product of nonintentional or *causal* processes of emergence, manifested in the automatic process of photographic registration, while the latter is entirely intentional, the product of the deliberate imaginative efforts of the artist. Rodowick is, we could say, as so many theorists of the digital cinema are, "obsessed" by physical phenomena, by the material and physical differences between the digital and the analog cinema, which he presumes are the sources of differing ontologies. If, though, as Searle argues, "speech" as an intentional act of representation cannot be reduced to "sound waves," to be contrasted, for instance, with writing, reduced to the material facts of ink and paper, as an "ontological" distinction, as the basis for an ontological distinction between the spoken and the written (as, for instance, Jacques Derrida so famously proposed), nor can the physical differences between the analog and digital be the basis for distinguishing in any ontological sense between two kinds of cinema, two kinds of cinematic representation. It is also false, in this respect, to suggest, as Rodowick does, that the digital is more amenable, somehow, to the creation of counterfactual, *i.e.*, imaginative, or *fictional* representations. The fictional is logically only one aspect of intentional representation, as Searle argues. "In fictional discourse," he says, "we have a series of pretended (as if, make believe) speech acts, usually pretended assertives, and the fact that the speech act is only pretended breaks the

60 Ibid.; my emphasis.

61 Ibid., 15–16.

world-to-world commitments of the normal assertive."[62] The fictionality of any cinematic utterance, then, whether created photographically, or through any other means, even digital means, is not an effect of those means, a material effect, or formal effect, but an effect of the logic of intention, of the purpose of the "act" of utterance. Photographs, regardless of their causal origins in "reality," as recordings, do not, *in themselves*, limit or attenuate the fictional (or deceptive) purposes to which they may be so easily put, fiction (and deception) being an effect of a logic of utterance, of speech act, rather than an effect of some particular material.

For most theorists of the digital cinema, though, the ontological difference between it and the so-called analog cinema is reduced to a difference in the physical substances of the utterance. Indeed, the physical basis of the analog cinema, supposedly reducible to the process of photographic registration, is in fact thought to (partially) exclude that cinema from the very realm of the intentional, of the representational. Indeed, Rodowick says explicitly that he has "followed Stanley Cavell in insisting that photographic capture is not representation."[63] In its origins in "automatic analogical causation," that is, the photograph, as the basis of film form, was "*necessarily* tied to physically existing spaces and times, even though captured elements may be recombined to produce imaginative worlds and counterfactual senses."[64] As Searle explains, though, any "world-to-world" relation is established by the logic of utterance, not by the putative ontological quality of any substance, which would have to be manipulated or altered, to change the nature of the substance, as Rodowick implies, but by altering the logical structure of the "assertion."

The question of representation, though, for Rodowick is reduced to a matter of substance, to, we may say, *matter*. In this respect, he thinks of photography, as a material substance, as a kind of matter, as imbued with an inherent ontological quality. As Rodowick explains earlier, in what can be seen to be in distinct contrast to Searle's analysis, "Cavell's use of the

62 Ibid., 18.

63 Rodowick, *The Virtual Life of Film*, 106. It is worth noting that Roger Scruton reaches a similar conclusion, on the basis of which, however, he questions the aesthetic value of photography and cinema. "Photography is not representation; nor is it representation when it used in the cinema." On the basis of such an assumption, Scruton argues that the cinema derives its aesthetic value (if it has any) from other sources, such as dramatic performance, or the importation of painterly techniques, that indeed it must overcome its essentially non-representative origins, if it is to be art, the chances of which are, Scruton suggests, typically rather slim. See Roger Scruton, "Photography and Representation," in *The Aesthetic Understanding: Essays in the Philosophy of Art and Culture* (Manchester: Carcanet Press, 1983), 122.

64 Rodowick, *The Virtual Life of Film*, 106–7; emphasis in original.

terms 'ontology' and 'reality' in relation to film have nothing to do with the correspondence of an image and its referent. Our sense of the 'reality' of film comes not through representation or even the representativeness of its projected images, but rather through the way in which this projecting world confronts us with our own metaphysical condition."[65] Indeed, Rodowick describes the cinema in its original "analog" or photographic constitution, as "a machine for metaphysics, with a distinct place in the complex history of skeptical thought."[66] Rodowick effectively wants to exclude the "photograph," which he argues was the basis of the "analog" cinema, from the very realm of representation, in order to place it in a different ontological category from the "digital" cinema, whose relation to reality is mediated precisely by intention, as opposed to the supposedly more direct and therefore metaphysically revelatory relation of the photograph.

The cinema, in its digital form, in such an argument, effectively becomes evidence for, or the manifestation of, a preferred theory or metaphysics, a philosophical skepticism, which proposes that our knowledge of the world is necessarily unreliable and unverifiable, that we can never truly experience "reality" directly, while the cinema, in its original analog or photographic form, is presented as somehow capable of revealing this philosophical fact by resisting the epistemological security of "representation." There is, in such a view, a duality characteristic of perception, such that there is an external world, from which we are inevitably separated, about which we have only an "internal" representation, which we create with our minds, but which is not reality itself. In this respect, two different things are posited: reality and the *representation* of reality, which are supposed to correspond to, respectively, an *objective* realm of existence, which we supposedly cannot experience or perceive directly, and a *subjective* realm, where what we perceive are only our "representations' of that reality—the familiar Kantian tones of which should be recognizable.[67] Against such a view, Searle proposes what he

65 Ibid., 72.

66 Ibid., 66.

67 John Hyman has explored the legacy of Kantian, and Hegelian, philosophy on matters of visual depiction. "Ever since Kant placed the problem at the center of the *Critique of Pure Reason*," Hyman says, "philosophers have struggled to define the extent to which the constitution of the human mind influences our knowledge and experience of the world." Whatever the virtues of such a philosophical view, though, he argues, it cannot be the basis for a theory of depiction. Yet, as he notes, most such theories have effectively argued, as Rodowick does, that pictures, depictions, are fundamentally manifestations of a favoured philosophical theory. "The extraordinary claim," he says, "in Hegel's *Lectures on Fine Art*, that the essential principle of painting is subjectivity of mind, is accepted by the majority of philosophers today." Against an effectively "causal" theory of depiction, based on a causal theory of perception, Hyman offers a

calls "a noninferential, that is, a naïve realist, version of the causal theory of perception, according to which we are not dealing with two things one of which is the evidence for the other, but rather we perceive only one thing and in so doing have a perceptual experience."[68]

I am convinced by Searle's realism, as opposed to this classical form of skepticism, which we can trace back at least to Descartes and to Kant. For my purposes here, though, it is not important to refute Rodowick's skepticism, but rather his even less tenable claim that "the cinema," as an artistic form, as a mode of expression, is in fact capable of avoiding the very condition of the representational, and thereby reveals the truth of skepticism, by example, so to speak, by manifesting the very subjective experience that skepticism posits, rather than by argument, which, as far as I am concerned, is the only mode by which skepticism, or any other preferred philosophical theory, can be proffered. Of course, a filmmaker may, in a particular film, and as part of the fiction that they present, offer a skeptical theory of reality, as say, the Wachowskis do in *The Matrix* (1999), but another filmmaker may just as easily offer a naive realism, comparable to Searle's view, as, I think, for instance, Eric Rohmer does in many of his films—which are so often just about what we really see, or about seeing things as they (really) are. The problem of skepticism, that is, and arguments for and against it, may be the *subject* of a film, may provide the content of a film, but cannot be corroborated by the simple fact of film's physical or material existence as a (photographic) medium. This, indeed, is the danger of most "ontological" accounts of the cinema, which posit a significance to any particular physical material, the "analog" process of photography against the "digital" process of simulation, and derive from it an identity, or what can only be described as a *revelatory* capacity, beyond the strictly representational and intentional, beyond any particular intentional representation.[69] Indeed, by describing the cinema (at least in its analog form) as a "machine for

renovated theory of depiction as resemblance. See John Hyman, *The Objective Eye: Color, Form, and Reality in the Theory of Art* (Chicago: University of Chicago Press, 2006), xvii, xviii.

68 Searle, *Intentionality*, 74. Searle has more recently developed his rebuttal of skepticism, which he characterizes as basically Kantian, and which he argues is based on a quite colossal philosophical mistake, which has, he says, "overwhelmed" the philosophical tradition. It is "the mistake of supposing that we never directly perceive objects and states of affairs in the world, but directly perceive only our subjective experiences. The mistake," he goes on, "has many different names, among them Descartes, Locke, Berkeley, Leibniz, Spinoza, Hume, and Kant. After Kant it gets worse." See John Searle, *Seeing Things as They Are: A Theory of Perception* (Oxford: Oxford University Press, 2015), 11.

69 On the history of the "revelatory" quality of much film theory, see Malcolm Turvey, *Doubting Vision: Film and the Revelationist Tradition* (Oxford: Oxford University Press, 2008).

metaphysics," Rodowick is subscribing to a longstanding tendency in the theory of art that Jean-Marie Schaeffer has described as the "speculative theory," according to which art is, or is potentially, or ideally, the source of a kind of philosophical insight or knowledge, of revelation, of an "ecstatic" knowledge, manifested in the form of a search for the "essence" of art, insofar as the primary question of the theory of art is supposed to be, "What is art?" The answer, very often, has been that art is a particular (privileged) mode of cognition, or the means by which we achieve a certain, unique "cognitive status." As Schaeffer says: "we are told that art is an ecstatic knowledge, the revelation of ultimate truths, ... a transcendental experience, ... the presentation of the unrepresentable, the event or occurrence of Being; and so on."[70] Such a theory, he argues, effectively implies a "sacralization of art, which is contrasted, as an ontological mode of knowledge, to other human activities, which are seen as alienated, deficient, or inauthentic."[71]

I would add, and as in the case of much film theory, that there is a distinction made between art that realizes such an ontological goal, and art that fails to do so. The analog cinema is often presented as the form of cinema that had achieved such status, often thought to have achieved it to a degree unprecedented in the history of art, while the digital is a necessarily lessened form, which is incapable of ecstatic revelation (or, as Rodowick in fact argues, must find new means of achieving such a status). In any case, and as Schaeffer argues, like Searle (whom he does not mention), what exponents of such a theory of art may or may not know or acknowledge, is that "this thesis presupposes a theory of Being: if art is ecstatic knowledge, this is so because there are two kinds of reality, the apparent one to which we have access through our senses and reasoning intellect, and the hidden one that reveals itself only to art (and perhaps to philosophy)."[72] Yet, as Schaeffer says, such a theory is one that would have to be argued for, that cannot be understood to be revealed, somehow, by the contemplation of art, or of any particular, privileged kind of art. On the basis of such an assumption, though, a specific discourse on the arts emerges, whose "task is to provide a philosophical legitimation of the ontological cognitive function of art. What this amounts to is the claim that the arts and art works have to legitimate themselves philosophically."[73] The works of art, or the specific art forms, that are privileged in such theory, as

70 Jean-Marie Schaeffer, *Art of the Modern Age: Philosophy of Art from Kant to Heidegger*, trans. Steven Rendall (Princeton: Princeton University Press, 2000), 6.

71 Ibid.

72 Ibid.

73 Ibid., 7.

Schaeffer says, for such theorists, are those that are "in conformity with their postulated philosophical 'essence': whence precisely the necessity for artists to envisage their works as answers to the question, 'What is art?'—understood as a question about its legitimacy. Thus, the circle is completed: the search for the essence is in fact a search for philosophical legitimacy."[74]

The crisis of the digital cinema, in this respect, and the crisis of digital editing, as Fischer describes it, and in the form of Rodowick's argument and of the arguments of many film theorists, is in fact a crisis of the theoretical discourse of the cinema, a crisis of film theory, which has long struggled to overcome some deeply ingrained and longstanding assumptions about art itself, applied to the cinema as an art, which has so often been reduced to the "photographic," against which the intentional efforts of the film artist are supposed to have been directed. Understood as essentially photographic, the cinema was thought necessarily to place limitations on artistic intention, which determined the specific shape and unique value of works of cinematic art, to emphasize its ontological cognitive function. Most theories of film tend to discover in the cinema, in film art, evidence of a favored view on art or a preferred style, as works of film art are read programmatically as manifestations of a favored metaphysical outlook, a worldview, or an aesthetic predilection—as a philosophy, or as a theory. As Schaeffer argues, though, "art works cannot be reduced to plans, nor a fortiori to theories or to movements or to schools."[75] Art is instead, he says, what arises from the necessarily diverse intentional efforts of artists. Yet a great deal of art theory, he argues, subscribes to just such a reductionist argument, in as much as theory is so often in the form of a search for an artistic essence. "But fundamentally," Schaeffer says, "the essentialist quest makes no sense; art is not an object endowed with an internal essence; like every intentional object it is (becomes) what people make of it—and they make the most diverse things of it."[76]

***

The effectively essentialist account of film art, and the concept of intention, as one kind of aesthetic value, and as manifested in a particular technique, and in an agonistic relation to a putatively nonintentional aesthetic value, derived from a different part of the technical apparatus of the cinema, is an

74 Ibid.
75 Ibid., 4.
76 Ibid., 6.

idea that has developed historically in the theoretical discourse on film. The value of editing as a technique has typically been thought to derive from the particular material constitution of the cinema, so that such value is understood to change as a result of the changes to that constitution, insofar as the basic distinction between intentional and nonintentional forces in the cinema is dissolved. Indeed, as Rodowick argues, "the long privilege of the analog image and the technology of analog image production have been almost completely replaced by digital simulations and digital processes."[77] Among the "processes" that have been replaced, or rendered in digital form, for Rodowick, editing is perhaps the most important, which, in the new form of digital montage, exploits the full potential of digital simulation. Rodowick argues that the most significant effect of the so-called "digital revolution" in the cinema, the incorporation of digital or computer technologies into the apparatus of filmmaking, is the reduction in the value of editing as a technique, manifested, paradoxically, through the triumph of editing as digital montage—the "reduction" understood as the result of the removal of the material constraints to editing. That constraint, understood as an inherent aesthetic value, he argues, had been derived from the specific material quality of the cinema, from its original technological basis, which has been replaced by a new technology. As Rodowick says, "for the first time in the history of film theory the photographic process is challenged as the basis of cinematic representation."[78] The consequence of this is most obvious, he argues, through a consideration of the transformation of editing, the advent of an aesthetics of digital montage, as the "analog" process of photographic recording is replaced by "digital" simulation.

Most specifically, the basic material of the cinema, or the integral unit, the *shot*, the product of the original technological process of photographic recording, is transformed. "The 'analog' shot," he says, "defines discrete spatial wholes, and thus film editing entails combination of these wholes."[79] But this is what has changed, Rodowick argues, with significant aesthetic and ontological consequences. "Where film editing may be characterized as the addition of spatially discrete wholes," he says, "the modularity and discreteness of the numerical image allows it to be separated into or combined with a potentially infinite number of digital events that can be manipulated in a great variety of ways."[80] The digital cinema, Rodowick

77 Rodowick, *The Virtual Life of Film*, 9.
78 Ibid., 9.
79 Ibid., 170.
80 Ibid.

argues, in contrast to the photographic or "analog" cinema that it is supposed to have replaced, is characterized "by an aesthetic of postproduction, highly subject to computational processes and the imaginative intentions of its authors."[81] For Lev Manovich, by contrast, the advent of a digital cinema, and of a newly expanded form of what he calls "spatial montage," is to be welcomed, insofar as "film obtains a plasticity that was previously only possible in painting or animation."[82] As Manovich insists, as the shot is opened up for spatial manipulation as well as temporal arrangement, as editing is expanded by digital technology, "directions that were closed off at the turn of the century when cinema came to dominate the modern moving image culture are now again beginning to be explored."[83]

The recent debates about digital editing in the cinema have revealed some of the most significant underlying assumptions in much film theory, about the relation between technology and technique, the technical and the aesthetic, about the role of the film artist, and about the vexed question of intention. The most basic assumption, as expressed here by Rodowick, and as I have already considered, is that the digital cinema is the triumph of intention and imagination, manifested most acutely in the expanded and extended powers of editing in the form of digital montage. Manovich shares the same basic assumption, that, in its original form, the cinema consisted largely of "unmodified photographic recordings of real events that took place in real, physical space."[84] For Manovich, though, this was a limitation to be overcome, ultimately by digitization, finally allowing for full intentional control, manifested specifically in an expanded form of editing. For Rodowick, editing had been properly constrained, or limited, by the material quality of the film image itself, and that the primary source of aesthetic value in the cinema was the essentially nonintentional process of photographic recording.

Yet this is, in both cases, to confuse the concepts of the aesthetic and the technical, which are, in such accounts, distinguished and separated. Both Manovich and Rodowick explicitly contrast the intentions of the film artist with an autonomous technical apparatus. For Manovich, artistic intention has finally been released, having been limited by the "indexical" quality of the photographic recording, freed now by digital montage; for Rodowick, artistic intention and imagination has been dangerously unloosed. Such a

81 Ibid., 173–74.
82 Manovich, *The Language of New Media*, 301.
83 Ibid., 308.
84 Ibid., 294.

distinction, though, in either case, between, on the one hand, an aesthetics of the cinema, and, on the other, a technical apparatus that either limits or allows the expression of such an aesthetic, and the contrasts established on the basis of such a distinction, cannot be logically maintained. A film, like any other work of art, consists necessarily, as Gérard Genette has persuasively argued, of two properties, which are precisely the aesthetic *and* the technical, which are necessarily and intimately related. "An artwork," as Genette says, "is an object that is simultaneously (and intentionally) aesthetic and technical; its technical properties are aesthetically relevant to the extent that they shape our appreciation of this work."[85] We may, of course, appreciate a work of art without any consideration of its technical origins, of the technical means employed to create it, or any consideration of the ways in which those means were employed. Indeed, by simply looking at a work of art, on the basis of our direct perception of it, we may well not be able to determine its origins in any technical process or undertaking, or even be interested in trying to discover them. Nevertheless, we may still be able to, and are free to, render an aesthetic judgment. As Genette says, "a primary aesthetic judgment based solely on perceptual data is, in a sense, perfectly legitimate."[86] Still, some more knowledge about the work's technical aspects, Genette says, potentially "enriches" one's relation to it, what he describes just as the "aesthetic relation," contributing to the experience of the work "as artistic object."[87] Understood as such, as an "artistic object," the cinema is necessarily the site of the intentional use of technical affordances to create aesthetic effects. Whether such affordances are "analog" or "digital," whether they are photographic recordings or animated simulations, whether created by a camera or through the use of special effects, is, in itself, not the source of specific value. It is important to understand the changes to the technology of the cinema, and to the techniques of editing, but according to a more coherent account of the relation between the technical and the aesthetic.

***

The origins of the philosophical debate about the digital cinema, and about digital editing, can be found in the original debates about editing itself. Editing has long been described as the specific or most important means by

85 Gérard Genette, *The Aesthetic Relation*, trans. G. M. Goshgarian (Ithaca: Cornell University Press, 1999), 151.
86 Ibid., 153.
87 Ibid.

which artistic intention is realized in the cinema, in contrast to the mutely causal operation of the camera, understood as ultimately little more than a mechanical recording device. Indeed, editing has seemed like the basic aesthetic prerequisite for film art, as the technique necessary for film to *be* art. As Lucy Fischer argues, the new digital editing technology, emphasizing and strengthening editing *as* a technique, restages the traditional debates about the relative value of editing, as the means by which the recordings of "reality" would be modified, to suit artistic intentions, or would function to reveal that reality, by subordinating such intention to the causal force of the recording itself. Such figures as Rudolf Arnheim, she notes, saw editing as one of the most important formal techniques for film art, with which it would transcend its origins in the process of merely mechanically recording reality to achieve a genuinely aesthetic status. If, as Fischer says, Arnheim "sees the discourse of editing as entirely unconstrained (as compared to the rules of the world's space-time axis)," capable of creating jumps or "jerks" in space and time that are impossible in "reality," he nevertheless, at the same time, "understands that montage can be used to create a sense of reality on-screen."[88] As Arnheim says, in a passage quoted by Fischer, while editing provides considerable artistic "freedom" to the filmmaker, that freedom "is usually restricted in that the subject of a film is an account of some action, and a certain logical unity of time and space must be observed into which the various scenes are fitted."[89] Still, the other possibilities of editing, arguably more important for Arnheim, and as explored most famously and doggedly by the Soviet-era filmmakers and "montage" theorists whom he discusses, are those created when, as Arnheim says, "shots are associated by montage whose connection is not realistic but conceptual or poetic."[90] Indeed, for Arnheim, editing could seem to be the obvious source of aesthetic value in the cinema, as the montage theorists and filmmakers like Lev Kuleshov and V. I. Pudovkin, and especially Sergei Eisenstein, claimed. "It is easy to understand," as Arnheim said, "how someone could come to think that it was first with editing that film as an art began."[91] It was

88 Fischer, "Film Editing," 66.

89 Quoted in ibid., 67. See Rudolf Arnheim, *Film as Art* (Berkeley: University of California Press, 1957), 21.

90 Quoted in Fischer, "Film Editing," 67. See Arnheim, *Film as Art*, 89.

91 My translation of: "Man versteht aber leicht, wie jemand darauf kommt, dass erst mit der Montage die Filmkunst anfange." See Rudolf Arnheim, *Film als Kunst* (Suhrkamp, 2002 [1932]), 95. This is more literal than the standard English translation, with its vaguely Freudian suggestion, which is: "It can be easily seen, however, why montage might be thought of as the royal road to film art." See Arnheim, *Film as Art*, 87. I have also translated the German noun "*Montage*" more

the montage theorists, who argued for the value of editing and offered an artistic theory of editing practice, he says, who "first realized the artistic potentialities of montage, ... who first made an attempt to define its principles systematically."[92] In fact, Arnheim thought that the montage theorists exaggerated the effects and the importance of editing, admitting that they "often carried their enthusiasm for it too far."[93] Yet he could understand the distinction that was at the basis of the theory. "The single film image," Arnheim says, "after all, originates in a recording process, in which one intervenes only as a guide, but which, regarded superficially, otherwise merely reproduces nature."[94] With editing, by contrast, Arnheim says, "one takes a hand in the process," breaking up the sequence of events, creating new spatial and temporal relations, in, he says, what "looks much more like a tangibly creative and formative process."[95] The importance of the technique cannot be underestimated, he says. "With editing," Arnheim argues, "the film artist has a first-class means in hand, to shape, interpret, and deepen the process of reality that he had recorded."[96]

If Arnheim was willing to accuse the montage theorists of taking their enthusiasm for editing too far, others argued that it should in fact be quite strictly limited as a technique, so that it would not interfere with the more basically realistic value of the cinema. A countertendency emerged in film theory, as Fischer notes, which emphasized the need to limit or restrain the techniques of editing, to serve primarily the powerfully realistic quality of cinematic representation, based in the more essential fact of recording, in the process of photographic registration. Siegfried Kracauer distinguished editing as a "technical property" of film, as opposed to the more fundamental "basic property" of photography, but he was still willing to say that of such properties, "the most general and indispensable is editing. It serves to establish a meaningful continuity of shots and is

simply as "editing," according to normal usage in German, now and at the time, and to avoid confusion with the Soviet theory that Arnheim considers. It is not stated who the translator was of this English edition, and could well have been Arnheim himself. In an earlier translation, from 1933, credited to L. Marie Sieveking and Ian F. D. Morrow, the passage is rendered in somewhat more literal, if not, to my mind, more satisfactory form: "It is, however, quite intelligible that anyone should consider that the art of film begins only with *montage*." See Rudolf Arnheim, *Film* (London: Faber and Faber, 1933), 95.

92 Arnheim, *Film as Art*, 87.

93 Ibid.

94 Ibid., 87–88; I have slightly modified the translation for clarity; cf. *Film als Kunst*, 95.

95 Ibid., 88.

96 Ibid., 89; translation modified and expanded; cf. *Film als Kunst*, 97.

therefore unthinkable in photography."[97] Yet, and while the technique of editing in the cinema may in fact be what importantly distinguishes the cinema from photography, by transcending the specific limitations of *still* photography, it is nevertheless from photography that the cinema derives its primary aesthetic value, Kracauer argues, insisting that the cinema is an art (a unique art), insofar as "its basic properties are identical with the properties of photography."[98] Editing, as a technique, then, is subordinated to the more "basically" photographic properties of the cinema. "The interest lies not with editing itself," he says, "regardless of the purposes it serves, but with editing as a means of implementing—or defying, which amounts to the same thing—such potentialities of the medium as are in accordance with its substantive characteristics."[99] In its primary characteristics, of course, for Kracauer, the cinema is substantively photographic. "Film," as he says, in his famous formula, "is uniquely equipped to record and reveal physical reality and, hence, gravitates toward it."[100] Beyond the basic qualities, as Kracauer admits, other techniques became important. There is, as he says, a "formative tendency" in the cinema. "As cinematic techniques developed," he says, "films increasingly drew on camera mobility and editing devices to deliver their messages."[101] As he implies here, though, the cinema is in a nobler business than merely "delivering messages." As he states blankly, "films may claim aesthetic validity if they build from their basic properties; like photographs, that is, they must record and reveal physical reality."[102] As he insists, a film is "cinematic" if and only if it "acknowledges the basic aesthetic principle," arguing that "the cinematic approach materializes in all films which follow the realistic tendency."[103]

The most famous critic of editing as a technique, of course, is André Bazin. While Bazin was keen to describe the "limitations" of editing, he was also acutely aware of and willing to extol its "virtues." If, in Eisenstein's film theory and filmmaking especially, a too great emphasis was placed on editing, and if editing was very often used in the cinema since Eisenstein's time in careless or flagrant fashion, to an excessive degree, it was certainly

97 Siegfried Kracauer, *Theory of Film: The Redemption of Physical Reality* (Princeton: Princeton University Press, 1960), 29.
98 Ibid., 28.
99 Ibid., 29. The passage is also quoted in Fischer, "Film Editing," but she provides an incorrect page reference.
100 Ibid., 28.
101 Ibid., 34.
102 Ibid., 37.
103 Ibid., 38.

not to be simply ruled out as a technique. "It is in no sense a question of being obliged to revert to a single-shot sequence," Bazin says, after criticizing some examples of what he saw as improper editing, "or of giving up resourceful ways of expressing things or convenient ways of varying the shots."[104] For Bazin, it is important to insist that "the abstract quality of editing is not absolute, at least not psychologically."[105] As viewers, he says, "we do not see editing, in its primordial simplicity, as artifice."[106] Indeed, he says, as audiences had grown more familiar with film and with the techniques of the cinema, they seemed to recede from view *as* techniques, especially, perhaps, editing, which does not interfere with the viewer's ability to become engaged with if not immersed in the representations being presented. Yet this does not mean, he argues, that editing is entirely legitimate, and that it should be used carelessly, even if it can be used effectively, to the point that we do not even really see it, but rather experience a sequence of shots separated by cuts as a spatially and temporally integral whole, like reality itself. Still, we can see it if we look, and draw what he says is the proper aesthetic conclusion. As he says, "a large proportion of a movie audience today, if asked to concentrate for a moment, would be able to distinguish 'real' scenes from those suggested by editing alone."[107] Used in a properly limited manner, that is, editing does not interfere with the inherently realist quality of cinematic representation, but recedes, becoming, as such editing technique is commonly called, effectively "invisible." What is thus retained is a basic integrity of the experience that corresponds, in an important way, to the integrity of our basic perceptual and conceptual experience of reality itself. As Bazin says, "the imaginary world shown on the screen must have the spatial density of reality [*la densité spatiale du réel*]. Editing can be used in such cases only within precise limits, on pain [*sous peine*] of threatening

104 André Bazin, "The Virtues and Limitations of Montage," in *What Is Cinema?*, trans. Hugh Gray (Berkeley: University of California Press, 1967), 50. The title of the essay in French, "Montage interdit," is rather more strident, and is translated more literally by Timothy Barnard. See André Bazin, "Editing Prohibited," in *What Is Cinema?*, trans. Timothy Barnard (Montreal: Caboose, 2009), 73–86. In subsequent quotations, I will use the Barnard translation, noting his title for the essay, which is in these cases more literal and accurate.

105 Bazin, "Editing Prohibited," 77. Barnard here translates the French term "*montage*," used by Bazin, more properly, I think, as "editing." Cf. Hugh Gray's translation, 45, which retains "montage." In French, Bazin says, "la nature abstrait du montage n'est pas absolue, du moins psychologiquement." See André Bazin, "Montage interdit," in *Qu'est-ce que le cinéma?* (Paris: Editions du Cerf, 2001 [1958]), 54.

106 Bazin, "Editing Prohibited," 77.

107 Ibid., 78. The adjective Bazin uses in French, "*réelles*," is presented in the original in qualifying quotation marks, which Barnard reproduces, but Gray does not.

the very ontology of the cinematic fable."[108] Again, it is not the case that Bazin proposes a comprehensive prohibition against editing, a complete interdiction. Editing is "prohibited" only when it threatens the more basic aesthetic value of realism. The long take, the single-shot sequence [*plan-séquence*], are not, in themselves, the only viable alternative techniques, but are to be chosen only when editing would otherwise become too excessive. The techniques of editing are to be understood as the "expressive resources [*ressources expressives*]" that they are, but they are not to be elevated to the principle of "montage," as proposed by Eisenstein and others. "We are not concerned here," Bazin says, "with the form but rather with the nature of the story-telling, or more precisely with a certain interdependence of nature and form."[109]

In many of the most significant accounts of digital editing, film theorists have returned to the debates of such "classical" film theory, either to the arguments of the "formalists," like Arnheim, and especially Eisenstein and the other "montage" theorists, to defend the expansion of the powers of editing in digital form, or, and in fact more commonly, to the arguments of the "realists" like Kracauer, but especially Bazin, to theorists of a photographic realism who find its source not in any particular technique but rather as something more fundamental, that "emanates" from the automatic and independent process of photographic recording, which technique can only suppress or emphasize. "Comparing computer-generated images with film," Rodowick argues, "reaffirms that photography's principal powers are those of analogy and indexicality."[110] He describes the photograph as "a receptive substance, literally etched or sculpted by light forming a mold of the object's reflected image," which grants it "both spatial and temporal powers that reinforce photography's designative function with an existential claim."[111]

108 Ibid., 80–81. Cf. Gray's translation, 48.
109 Ibid., 82. I have slightly modified Barnard's translation here. In French, Bazin says: "Ces présentes remarques ne portent pas sur la forme mais sure la nature du récit ou plus exactement sur certaines interdépendances de la nature et de la forme." Bazin, "Montage interdit," 59. There is admittedly some ambiguity in the phrasing. The term "*nature*" is first applied to qualify the noun "*récit*," or "narrative," what Barnard translates as "story-telling." (Gray, somewhat literally but entirely unsatisfactorily, translates "*récit*" as "the recital of events.") In any case, the noun is repeated in Barnard's translation, on the assumption, it seems, that it is still functioning as a qualifying term, still implying the presence of the noun "*récit*," or "story-telling." I think that Bazin is in fact making a more principled claim here, precisely about the interdependence of "nature and form," as a constituent aspect of the cinema as a medium, both words now rendered as substantive nouns in the concluding phrase of the sentence.
110 Rodowick, *The Virtual Life of Film*, 9.
111 Ibid., 9–10.

He cites Roland Barthes, for whom, as Rodowick explains, "photography is an 'emanation of the referent,' that testifies *ça-a-été*: this thing was; it has a spatial existence that endured in time."[112] He admits that digital processes are capable of producing images that are astonishingly "photographic," with the same quality and detail, that, as he says, "digital cameras, or even 'virtual' cameras creating wholly synthesized spaces on computers, are still based on the same optical geometry as traditional cameras and rely on the same historically and culturally evolved mathematics of depth and light rendering descended from *perspectiva legittima*."[113] Indeed, he says that there is "much to be learned from the fact that photographic 'realism' remains the Holy Grail of digital imaging."[114] Nevertheless, and as his qualifying quotation marks suggest, "what remains absent from the process of digital representation is what thinkers like André Bazin and Roland Barthes held fundamental to the photographic image: its causal force as a literal spatial and temporal molding of the originating event, preserved in a physical material."[115]

With the new technology, it is argued, the techniques of editing have indeed been elevated to the principle of "montage," at the expense of the necessary interdependence of nature and form, the supposed basis of the cinema's "ontology." There is a relatively clear consensus on this matter in the domain of film theory, a common understanding about the

112 Ibid., 10. See Roland Barthes, *Camera Lucida*, trans. Richard Howard (New York: Hill and Wang, 1981), 80.

113 Rodowick, *The Virtual Life of Film*, 11

114 Ibid.

115 Ibid. Katherine Thomson-Jones has considered what she describes as the "general worry about the impoverishment of the image in the digital age," given that it consists only of discrete elements, of pixels, which can only approximate the basic continuity of "analog" images. She insists, though, that such worry is misplaced, arguing that "a digital image need not appear any less richly detailed or varied in color than an analog image." She cites Rodowick, who distinguishes between the photograph, as the exemplary analog image, characterized by what Rodowick describes as "isomorphic transcription," and the digital image, which is merely a "data output." This is, Thomson-Jones says, presented as an "ontological divide," what she notes William J. Mitchell has called a "'sudden and decisive rupture' in the history of photography," which is then, by both, "assumed to have aesthetic implications." Despite what Rodowick describes as the "discontinuities" in the digital image, he admits, as she notes, that "a digital image can simulate the look of a continuously produced analogical image." There is a contradiction here, though, as she points out. "Such an acknowledgement," she says, "would seem to work against any attempt to identify the aesthetic effects of pixelation." Katherine Thomson-Jones, *Image in the Making: Digital Innovation and the Visual Arts* (Oxford: Oxford University Press, 2021), 19, 20. See William J. Mitchell, *The Reconfigured Eye: Visual Truth in the Post-Photographic Era* (Cambridge, MA: MIT Press, 1994), 59.

effects of the recent technological changes, which are understood to have had profound consequences for the cinema *as* an art, altering its very identity, or what is so often called its "ontology." That identity, the artistic and aesthetic value of the cinema, manifested in specific techniques, is thought to have been derived from a particular technological configuration, with the photograph at its heart. The photographic image, which had effectively defined the cinema, granting it its particular significance as an art, has been replaced now with what is called the "digital image." With this change, it is widely argued, a new kind of cinema is supposed to have emerged, a "digital" cinema, supplanting its traditional form, now often described as "analog," emphasizing the distinction. Complementing this change, the basic technology of editing has been transformed, with what are thought to be the most significant consequences, the result of the radical expansion of the power of editing as a technique. Indeed, it is with the more expansive powers of editing, it is typically argued, that the full capacities of the new digital cinema are realized, with new and more powerful tools to modify the radically more plastic material. Rodowick has made this argument in perhaps most compelling and explicit form, arguing that the "digital image is more and more responsive to our imaginative intentions, and less and less anchored to the prior existence of things and people."[116] The more "responsive" quality of the digital image, contrasted with the materially resistant photographic image, is exploited through the new means of electronic computer editing, as the basis of a much more expansive aesthetics of digital montage, in contrast to the traditional art of film editing, which is supposed to have been importantly limited as a technique by the technology of photography.

While the cinema, that is, could be put to intentional use, to present representations, with traditional formal techniques of editing to link images in meaningful ways, but also other techniques of compositing such as superimposition and optical printing, which could in fact alter the spatial elements of the shot itself, to create "imaginative worlds," the origins of these worlds were ultimately in the nonintentional process of photographic recording, which was what "tied it to" and established the meaningful relation with a preexisting physical reality, the source ultimately of the cinema's unique realism. By contrast, "digital synthesis is only optionally tied to the physical world through its capacity to construct spatial semblance."[117] The emphasis in the cinema now, he argues, is on synthesis and construction,

116 Rodowick, *The Virtual Life of Film*, 86.
117 Ibid., 107.

the specific values of editing, which was the most significant technique for the transformation of the photographic material into the form of a "representation." The powers of editing had always been constrained by the material quality of the photographic image, but have now been expanded and extended, overcoming what are understood to have been the basic *limits* of editing, constraints placed upon it by what is generally thought to have been the raw material of cinematic representation. That raw material, the basic constitution of the cinema, the material substance of cinematic representation, is understood to have been transformed, with the effect that editing, as a technique, has been radically enlarged.

The primary consequence, Rodowick argues, the triumph of editing, is precisely the triumph of the power of *intention*, which had been limited in the cinema, constrained by what he says was the inherently *causal* nature of the photographic image. The advent of a digital cinema, and the expanded powers of digital editing, have collapsed the boundary between the two, creating what he describes as an "ontologically curious" hybrid phenomenon. Now, he says, the image is always "montage." While once there had been a clear distinction *between* image and montage, between the shot and the editing, the former placing important limitations on the later, that distinction, Rodowick argues, has been dissolved, so that the digital cinema is characterized by a thoroughgoing plasticity, or mutability—what we could call a comprehensive *editability*—reduced to eminently manipulable minimal units, digital bits or pixels, in contrast to what he says was the material integrity of the photographic image, the basic and indivisible unit of the shot, defined as a "spatial whole," as he puts it. "What is combined in 'montage,'" Rodowick says—that is, by "digital montage" as opposed to film editing—"is no longer just contiguous spatial wholes as blocks of duration, for any definable parameter of the image can be altered with respect to value and position."[118] Indeed, as he insists, "there may be 'takes,' in digital cinema, but there are no shots. Or rather," as he explains, "what was previously considered a shot has now become a highly variable element open to interactive manipulation at the most discrete levels."[119] In the place of the usual concept of the shot, effectively destroyed by digital technology, Rodowick describes what he calls the "digital event," which is, he explains, "any discrete alteration of image or sound data at whatever scale internal to the image. Elements may be added, subtracted, or refashioned interactively because the data components retain their separate, modular identities

118 Ibid., 169.
119 Ibid., 166

throughout the 'editing' process."[120] Now placed in qualifying quotation marks, "editing" is understood to have been subordinated to the processes of digital synthesis and compositing, to digital montage and to the creation of "digital events" rather than traditional sequences of integral shots. The effects of digital editing contribute to what he describes as "the curious ontology of digital worlds."[121] Such peculiarity or ontological curiosity is the effect of the triumph of one technique over another, of "the technology of compositing" over editing, which had been understood as "combining discrete spatial elements recorded only from physical spaces."[122] The original discreteness of the shot, though, as an irreducible spatial element, has been breached. The basic quality of the photographic image, as the basis of the shot, has been compromised. As Rodowick says, "in digital cinema there is no longer continuity in space and movement, but only montage or combination."[123]

***

For Rodowick, and for other digital theorists, the aesthetic value of the cinema is reducible to its technical origins, and it is thereby susceptible to any significant change to its technological basis, with what are thought to be "ontological" consequences. There is, of course, some kind of relation in works of art between the technical and the aesthetic. Indeed, Genette has argued that an account of the "technical procedures" that go into the making of a work of art must be part of any viable and informative theory of art, or of what he calls the "artistic function" performed by a work that is created, as he says, to "solicit" our aesthetic attention, or presented as a "candidate" for aesthetic appreciation. An understanding or awareness of the technical or technological origins of a work of art, he says, is a necessary and irreducible aspect of what he describes as the "aesthetic relation" that we can establish with it. In fact, he argues, the "aesthetic" cannot be separated from the "technical," the "universality" of which, he insists, is a logically necessary part of our experience of works of art. Insofar as a work performs an artistic function, that function is "par excellence, the locus of interaction between the aesthetic and the technical."[124] A theory of the cinema, of film as art, should be able to describe such "interaction," which Genette argues

120 Ibid., 167.
121 Ibid., 171.
122 Ibid., 172.
123 Ibid., 172–73.
124 Genette, *The Aesthetic Relation*, 155–56.

is the manifestation of the fundamentally *intentional* character of art. As he says, quite clearly and simply, "the specific and, therefore, defining feature of works of art is ... that they proceed from an aesthetic *intention*, and accordingly perform an aesthetic function, whereas other sorts of things can at best elicit a purely attentional aesthetic *effect*."[125] It is the technical aspect of works of art, which we experience as intentionally designed, that distinguishes them in a basic and important respect, as Genette argues, from, for example, natural phenomena, which may themselves be appreciated aesthetically, but only as an effect, though, of our choice to attend to them in this way. It is perfectly reasonable to describe a sunset as beautiful and to take pleasure in the view from a hilltop of a valley below, but the origins of such beauty are not the result of the intentional use of a technology. As Genette says, "natural objects, being fundamentally nonintentional products of chance and necessity, do not display features that one could describe, in the strong sense, as 'technical.'"[126]

According to such a distinction, works of film art are no different, I will insist, than any other kind of art, characterized by a "fundamentally intentional character." Yet, and as I have tried to show, a great deal of film theory tends to describe the cinema, in its original form, as *less* intentional than other kinds of art, and to argue that the "digital cinema" is *more* or even fully intentional, by contrast. The cinema—again, in its original or "analog" form—is thought to produce something more like what Genette describes as an "aesthetic *effect*," which is what he says we can respond to when we attend to natural objects or phenomena, when we say that a sunset or a snowflake or a flower is beautiful, if we take (as we may) aesthetic pleasure in such things. It is in this respect that they are, as he says, "attentional" aesthetic objects, as opposed to works of art, which are "intentional," inviting us just as *works*, designed for aesthetic judgment and appreciation. As he says, "this aesthetic relevance of technical data constitutes the essential feature of the special status of artworks."[127] The source of aesthetic significance in the cinema, though, is often thought to be more essentially attentional, and to be properly constrained in its intentional qualities—to be, indeed, more like a natural object than a traditional work of art. This, I have been arguing, is to make a basic conceptual error, but one that is common, and which provides the basis for most arguments about the digital cinema, and about the technology and technique of film editing, which is starkly contrasted with digital montage. It is to this contrast that I now turn.

125 Ibid., 1; emphases in the original.
126 Ibid., 155.
127 Ibid., 155.

3. The Technology and Technique of Film Editing

Abstract
In this chapter I consider the general formal problem of editing, arguing that it has been a particularly hard problem for film theorists. I survey several significant accounts of editing as an artistic technique, which emphasize it as a conceptual problem, and which have offered various explanations for the difficulty. The problem of editing, I argue, is best understood as an instance of the more general problem of the relation between technology and technique, between the technical and the aesthetic.

Keywords: Cinema, technology, technique, form, aesthetics, ontology

The basic technology of editing, the technical apparatus, has indeed changed utterly, dramatically transformed by the latest computer technology, with some notable consequences for editing as a technique. Explaining this transformation, though, describing what has changed and with what effects, is no simple matter, given the difficulty of describing editing *itself* as a technique, a difficulty that has been regularly commented on. Still, efforts have been made. Inevitably, though, in most analyses of editing, it is revealed to be a particular challenge, difficult to describe and explain even if it is presented as perhaps the most important and formally significant of the various cinematic techniques, the one in most need of explanation. Valerie Orpen, for instance, argues that editing is the source of cinema's basic expressive power, effectively offering a theory of editing as expression, but frankly admits to struggling when trying to analyze it. She suggests what the specific difficulty might be. "We know that editing exists," she writes, "even if it is often seamless."[1] The typical "seamlessness" of editing is a result, though, she suggests, of its actual intangibility as a technique. She observes

1 Valerie Orpen, *Film Editing: The Art of the Expressive* (London: Wallflower Press, 2003), 3.

Furstenau, M., *The Aesthetics of Digital Montage: Film Editing and Technological Change.* Taylor & Francis Group, 2024
DOI 10.5117/9789463722803_CH03

that, "strictly speaking, editing exists only *in relation to*, as a *counterpart* to, the shot. In other words, it is impossible to isolate editing, to analyse the cuts *per se*, that thin line, or 'switching,' that demarcates one shot from another."[2] Her goal, though, is to explain how this apparently intangible technique, whose effects are so seamless, is what indeed provides the cinema with its basic expressive power, describing editing as the "art of the expressive." Orpen ultimately admits, however, to the "impossibility of isolating editing, of disentangling it from other film-making aspects."[3] She suggests that "editing presupposes *mise-en-scène* in the fiction film," and argues that it is "far more useful to juxtapose *mise-en-scène* and editing than to separate them."[4] She concludes her book with an even more forceful declaration, insisting that "editing and *mise-en-scène* cannot, and should not, be separated."[5]

Editing is certainly more than just cutting, as Orpen argues, and cannot be understood except in relation to other film techniques, and not only *mise-en-scène*. Yet this makes it even more challenging to explain. This has led some to ignore editing, to give more emphasis to other more obvious and tangible techniques, in the face of the conceptual challenge that editing poses. As such a hard problem, editing has often been overlooked. Writing in 1980, Peter Wollen observed that, in contrast to the amount of attention paid to other aspects of cinematic technology, specifically to the facts of recording and of projection, the "history of the technique of editing … is rather neglected."[6] Orpen notes that this is still the case, observing that "there is a dearth of studies on editing, particularly in the sound film."[7] A further complicating factor, it could be argued, is that the peculiar intangibility, and consequent complexity, of the technique of editing is at apparent odds with its technical simplicity. "The material basis of film editing," as Noël Carroll has said, "is the cut, the physical joining of two shots. We can easily account for this process with a little chemistry," he says, noting only that, if we also consider "in-camera editing," then "we have to add some mechanics to our story."[8] That is all there really is to it, though, at least technically.

2 Ibid.; emphases in original.

3 Ibid., 115.

4 Ibid., 116.

5 Ibid., 119.

6 Peter Wollen, "Cinema and Technology: A Historical Overview," in *The Cinematic Apparatus*, eds. Teresa De Lauretis and Stephen Heath (London: Macmillan, 1980), 14.

7 Orpen, *Film Editing*, 14.

8 Noël Carroll, "Toward a Theory of Film Editing," in *Theorizing the Moving Image* (Cambridge: Cambridge University Press, 1996), 403. The essay was first published in *Millennium Film Journal* 3 (Winter/Spring 1979): 79–99.

However, as Carroll then declares, "editing involves more than chemistry and mechanics,"[9] and he goes on to elaborate a theory of editing as "communication." He offers one of the most explicit and detailed theoretical accounts of editing, which he analyses from a cognitivist perspective, as the primary technique for the creation of structures of inference and interpretation. As I shall consider later, though, he subsequently declared his effort a failure, as insufficient in the face of what turns out to be such a complex technique.

Characterized by a technical simplicity that obscures an aesthetic complexity, editing confounds many who have tried to explain it as a technique. As an essential aspect of the cinema, though, its simplicity seems also to have been the source of its stability. The technical apparatus of editing, unlike so many other aspects of filmmaking, has indeed been quite stable during the course of what has otherwise been an often tumultuous technological history in the cinema. Very little ever changed about the basic tools of editing, thus inviting less interest than other more technically dynamic aspects of the cinematic apparatus. The incorporation of synchronized sound into the cinema, for example, is widely regarded as one of the most profound and significant technological events, with consequent effects on most other technical aspects of filmmaking, including editing. Most of the technical challenges facing editors have been as a result of changes to other parts of the apparatus, and the advent of synchronized sound was arguably the most significant challenge, considered as such and most notably by André Bazin, as we shall see. It arguably became more difficult, but by no means impossible, to construct the kind of elaborate "montage" sequences of the so-called silent era, ushering in what could be described as a more tempered if not wholly conservative editing style in most cinemas around the world. In the end, though, and despite having to contend with the additional element of sound, and with the new technical positions in the editing profession required to contend with the fact of sound, editing remained basically the same as a technical and physical procedure, a process of cutting and reassembling pieces of film, only now with the addition of a soundtrack.

Still, there have been some quite significant changes to the technical apparatus of editing itself over the course of an otherwise remarkably stable technological history, though the stylistic or aesthetic effects have been difficult to determine with any certainty. Such changes can be seen, though, when looking at the tools and working conditions of early film editors. Wollen, for instance, describes revisiting the classic Soviet-era film by Dziga Vertov, *Man with a Movie Camera* (1929), which was edited by

9 Ibid.

Vertov's wife, Elizaveta Svilova, herself one of the major figures in the early years of Soviet montage theorizing and filmmaking.[10] This self-conscious avant-garde film is about the formal possibilities of the cinema itself, and specifically about the possibilities provided by editing, then being conceived as "montage," and presents the cinema, or cinematic art, as fundamentally the art of editing. As part of its formal strategy, the film includes scenes in the editing room itself, providing a glimpse, for later viewers, into the early history of editing, and of subsequent developments. As Wollen says, "with its shots of Svilova editing with a pair of scissors [I] was reminded ... of the changes which have taken place in editing equipment" (figs. 1–2).[11] Wollen does not in fact say what these changes were specifically, but at around that very time, when Vertov's film was made, in the late 1920s, the technology of editing was indeed undergoing a significant transformation, with new devices that had not yet reached the Soviet Union, given the political isolation of the new revolutionary country at the time. Most significantly, the scissors that Svilova is seen using had been replaced by more elaborate, dedicated editing devices, like the Moviola, introduced only a few years earlier, in 1924. Further developments would come later, when flatbed systems, like the Steenbeck and the K-E-M, were invented, becoming common by the 1960s. Once introduced, though, these various dedicated devices were used to perform the same basic action that had been performed with a simple pair of scissors, which was merely to cut strips of film into smaller pieces and join them together again. Still, such substantial changes as the introduction first of the Moviola and then of flatbed systems had effects. More minor changes, that may well have seemed otherwise negligible, could also be important, as Wollen suggests. "Even an apparently extraneous and trivial invention like Scotch tape," he says, "has had an enormous impact on editing technique."[12] Wollen does not say just what that impact was, however. Later in this book, though, I will in

10 Often overlooked in histories of editing, and even in histories of Russian and Soviet cinema, Svilova was a member of the so-called "Council of Three," the group of montage theorists and filmmakers, formed by Vertov (Denis Kaufman), with his wife and with his brother, the cinematographer Mikhail Kaufmann. Jay Leyda mentions her, describing the "boldness of Vertov and Svilova," which he argues is shown in the editing patterns of *Man with a Movie Camera*. See Jay Leyda, *Kino: A History of the Russian and Soviet Film* (Princeton: Princeton University Press, 1960), 251–52. Her role on the "Council," and her contributions to documentary filmmaking more generally, are considered in most detail by Christopher Penfold, "Elizaveta Svilova and Soviet Documentary Film." PhD diss., University of Southampton, 2013. https://eprints.soton.ac.uk/367302/1/C%2520Penfold%2520Thesis.pdf.

11 Wollen, "Cinema and Technology," 14.

12 Ibid., 14.

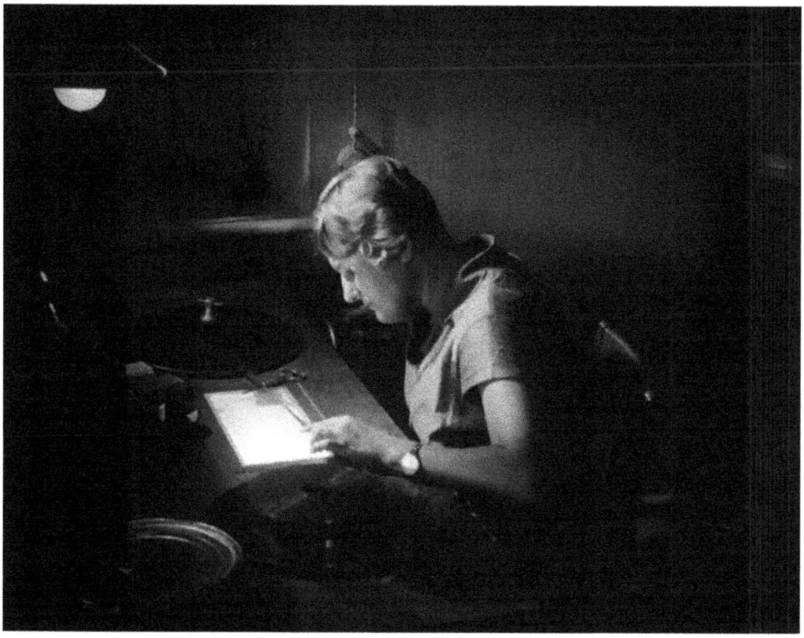

Fig. 1 Svilova with scissors. *Man with a Movie Camera* (Dziga Vertov, 1929).

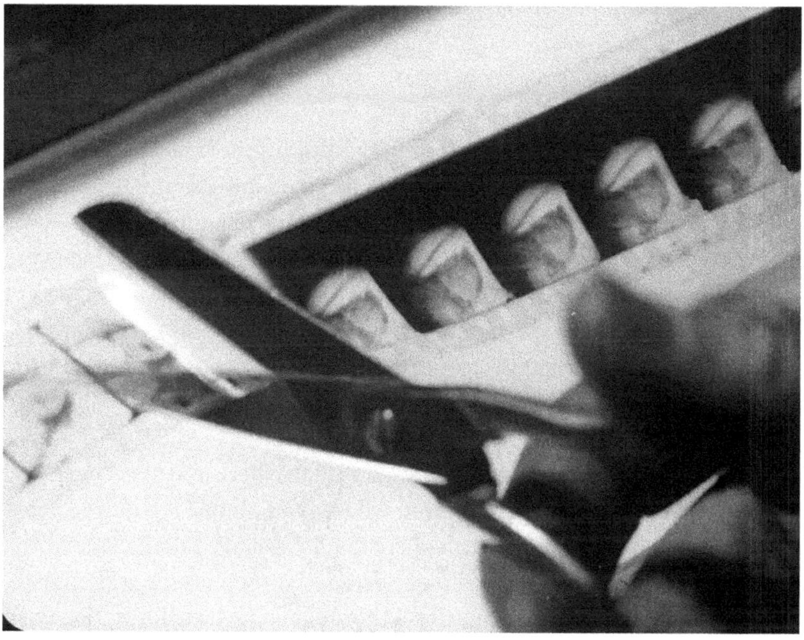

Fig. 2 Svilova cutting. *Man with a Movie Camera* (Dziga Vertov, 1929).

fact consider some of the effects that seem indeed to have resulted from the shift in editing from glue to tape, which were not inconsiderable, if not truly "enormous" as Wollen suggests.

Certainly, though, the scene of editing has changed in some specific and important respects since Svilova sat at her table with her pair of scissors and a pot of glue. Film editing as a technique begins with such humble, ordinary, and already existing technology, with a blade of some sort used to literally *cut* film, and some cement, or glue, and then tape, to join the pieces together again. Explaining how subsequent changes affect the artistic technique of editing, though, much less evaluating such changes as aesthetic advances, is difficult to do. Svilova's editing patterns, achieved in 1929 with the simplest tools, are arguably as complex and subtle, perhaps more so, than what we can see in subsequent decades of film history, even as the tools themselves became more sophisticated, as new technical components were incorporated, and even as, for example, the process of reassembling the pieces became much easier with tape rather than glue. The most recent changes have been even more significant. Films now are not even "cut" anymore; for the most part they are not even made out of "film" anymore. Yet it is not clear that the new digital systems are necessarily creating any new stylistic possibilities, beyond what, say, Svilova could already imagine with her scissors, her spools, sitting at the light-board with the strips of film being illuminated for her aesthetic consideration (fig. 3). In fact, what is remarkable, as we watch Svilova edit, comparing the scenes of editing in Vertov's film with the modern digital editing suite, are the significant similarities (fig. 4). The array of images, placed on the illuminated viewing board, which Svilova contemplates, considering the relations that she would like to establish between them, is essentially the same situation that every editor, every filmmaker, finds themselves in when approaching the task of constructing a film out of separate images (and sounds). The display surface now tends to be the computer screen, rather than the lightboard, but the *aesthetic* task, one is inclined to say, is effectively the same. The technology has changed, but the technique remains the same.

Yet there must be some differences, one feels; there must be something that distinguishes the task now from what it was almost a hundred years ago, given the very many technical changes that have taken place in the last two decades. After a very long period of only incremental change, there has now been what seems like a thoroughgoing transformation with digital technologies. There must be some consequences that have resulted from such changes. As even Wollen notes, though, it is not always easy to determine such effects, to trace the source of aesthetic development to

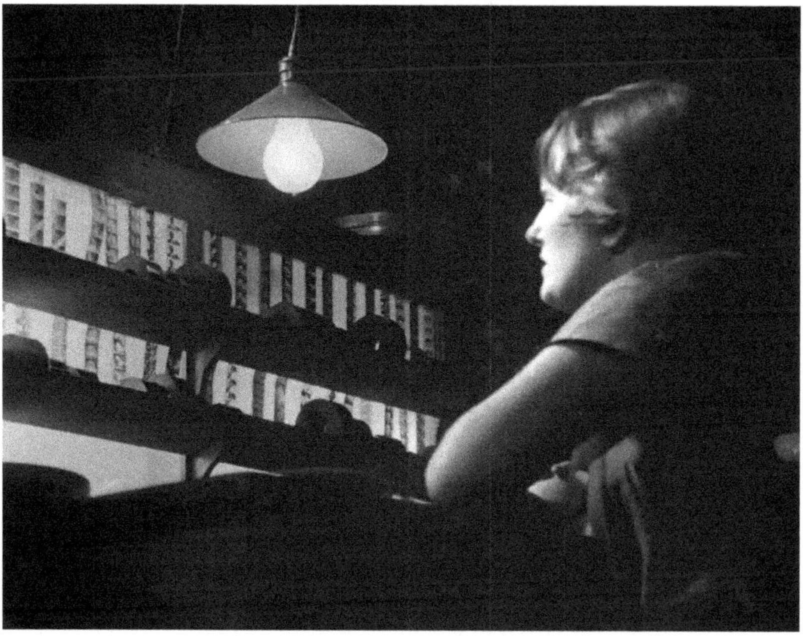

Fig. 3 Svilova at light-board. *Man with a Movie Camera* (Dziga Vertov, 1929).

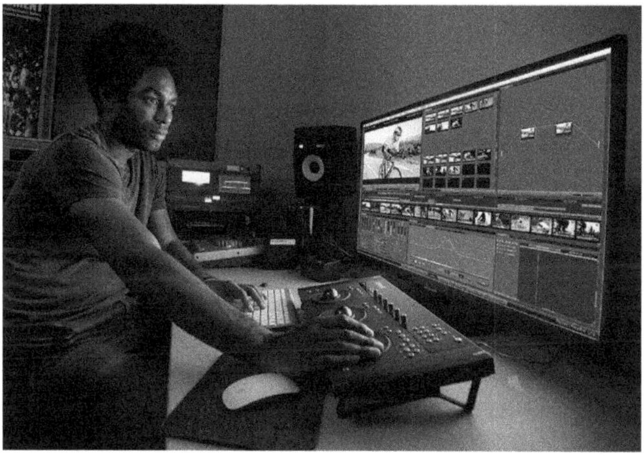

Fig. 4 A modern digital editing suite. Photograph by Ben Walker.
Source: www.sawvideo.com.

any specific technical change, given that "the technology of the cinema is not a unified whole, but is extremely heterogenous," covering such a wide variety of kinds of technical developments, in, for example, as he notes, "the fields of mechanics, optics, chemistry, and electronics," and including such disparate elements as "the Latham loop, Scotch tape, the densitometer, the zoom lens, magnetic tape, and so on and on (the list is long)."[13] Furthermore, any emphasis on any one of these various aspects will necessarily privilege it over any other, obscuring the in fact complex relations between them, as, it seems, Orpen is trying to suggest. Indeed, as Wollen says, this is what most theorists and critics had often done when considering the technological basis of the cinema as the putative source of particular techniques or stylistic effects, usually, though, without any explicit reason or justification. "In the past," as Wollen says, "theorists have tended to stress and even essentialise one or other area of technology at the expense of the others," such that the actual heterogeneity of cinematic technology is obscured, "reduced to one subset of determinations in a reductionist manner."[14] This, Wollen says, is the "myth of the cinema," conceived of as a supposedly homogenous phenomenon, every element securely linked to every other, creating a sense of completeness, of spectatorial plenitude, through a process that other theorists at the time were describing as "suture."[15] It is indeed a myth, though, Wollen insists, "which serves to efface the reality of production."[16] Such a myth, he says, is presented typically in the form of a putative *ontology*, the term most strongly associated with André Bazin, whom Wollen criticizes for what he says was his tendency to essentialize one specific aspect of cinematic technology, the techniques of recording, for Bazin, over any other.[17] The tendency persisted. Even within what he calls "avant-garde theory," the new theory of the time when he was writing, incorporating the insights of linguistics, semiotics, some Marxist theory, and psychoanalysis, which sought to overcome what were thought to be the limitations and prejudices of earlier, "classical" film theory, Wollen sees a similar propensity. He notes that, even if explicitly countering the theory of earlier figures, like Bazin,

13 Ibid., 20.

14 Ibid., 20–21.

15 The term "suture," based on a reading of Lacanian psychoanalysis, is usually traced to the essays published by Jean-Pierre Oudart, "La Suture, I et II," *Cahiers du cinéma*, nos. 211, 212 (April–May 1969): 36–39; 50–56. See also Daniel Dayan, "The Tutor-Code of Classical Cinema," *Film Quarterly* 28, no. 1 (Autumn 1974): 22–31. Central to the theory of suture was the process and effects of editing, as Robert Stam notes. For Oudart, Stam says, "the function of suture was to hide the fragmentation inherent in montage and thus conjure away the threat of the cut (evoking castration) and bind the spectator into the filmic discourse." See Robert Stam, *Film Theory: An Introduction* (Malden, MA: Blackwell, 2000), 137.

16 Wollen, "Cinema and Technology," 21.

17 Although Wollen in fact credits Bazin for the very concept of "myth," as precisely the kinds of stories about the history of the cinema that efface the technical realities of production. See André Bazin, "The Myth of Total Cinema," in *What Is Cinema?*, trans. Hugh Gray (Berkeley: University of California Press, 1967), 17–22.

criticized as "ontological," meant pejoratively as obscurely "metaphysical," in the new theory,

> there has been a parallel tendency to create an ontology, seeing means not as secreting a teleology, but as ends in themselves, as essences. The new ontology can easily be disguised as a materialism, since it seems to foreground means of production rather than "images." Yet, implicit within it is an assumption that the equipment used for making films is an essential bedrock rather than itself the product of a variety of historical determinations, at the interface where the economies of capital and libido interlock.[18]

While I will not pursue Wollen's own Marxist and psychoanalytically inflected account of such "economies," the difficulty that he describes, the very hard problem of describing the relation between technology and technique, between the material realities of film production and an aesthetics of the cinema, is arguably more pronounced now than it was then, even harder, and perhaps most acute when considering the technique of editing, already such a hard problem itself. Indeed, our ability to explain the consequences of technological change generally in recent decades have been especially challenged, as our world has been transformed by computers, by digital technologies, which Wollen was already beginning to glimpse in 1980, alert to the growing significance of electronics more generally. It was indeed with the development of electronic technologies, as Wollen in fact notes, in concert with the rise of computer processing powers, that the cinema was about to undergo what is arguably the greatest transformation in its technological history, the most significant since the technological irruption that was the invention of the cinema itself, the result of what was then an extraordinarily complex confluence of events in the history of technology. The literature on "the digital," on "new media," on the computer "revolution," is vast, but with a very general focus on the "digital cinema." My topic is narrower—the changes to the specific technical apparatus of editing—but it is necessarily a part of the much broader question of technology as it impinges on art, the very question indeed between technology and technique. As I have implied, given the relative stability of the technology of editing, it is perhaps where the effects of digitization are most obvious, if indeed more pronounced, now that such a sudden and sweeping change, arguably the greatest in

18 Ibid.

the whole history of editing, has taken place. Most accounts of the digital cinema tend in fact to isolate editing as that aspect where the change is thought to have been most acute, most significant, most consequential for the very aesthetic identity of the cinema. If, as Wollen suggests, the introduction of the use of something so apparently minor and insignificant as Scotch tape could have "an enormous impact on editing technique," then surely the comprehensive transformation of virtually all aspects of the basic technical apparatus for editing, its complete reconfiguration from a simple procedure of cutting strips of film to the total computerization of the process, will have had an even greater impact. Yet, of course, the story is not as simple as that.

***

If the effects of "digitization" in the cinema are often thought to be manifested most specifically, and most acutely, in the change to the technical apparatus for editing, this is largely because editing had for so long been thought of as perhaps the most consequential technique in the cinema. Despite its essentially invisible nature, its intangibility, allowing, as Orpen notes, for the kind of seamlessness that is characteristic of so much editing, the technique has long been thought to define the cinema as an art. Despite the considerable and well-attested difficulties of studying editing, there is a longstanding and widely held assumption that it is an especially significant technique, perhaps the most significant, the essential, determining technique that is the specific source of the cinema's aesthetic identity. While Orpen, like others, acknowledges the challenge that editing presents as an object of analysis, she insists that, as a distinct technique, it is indeed the basis of film art, the source, she argues, of its very power of expression. The implication, of course, is that an understanding of the cinema will thus depend upon an understanding of editing, just as the defining technique. This is the basic premise of the first and perhaps the most famous, certainly the most influential, theories of editing, which emerged in the politically and aesthetically charged context of the early Soviet Union, as Orpen notes. The first Soviet filmmakers, she says, "such as Vsevolod Pudovkin, Sergei Eisenstein and Lev Kuleshov, considered editing to be the creative force of film and the foundation of film art."[19] Orpen effectively accepts this central premise, concluding her study by declaring that, after all, and as the early Soviet theorists whom she had named had claimed, "editing is indeed the

19 Orpen, *Film Editing*, 1.

foundation of film art."[20] The view of these early theorists and filmmakers, essentially reiterated by Orpen, was that editing is an indispensable aspect of the cinema as an art form, indeed the technique that made possible its very emergence as an art. Any understanding of the cinema, then, will depend on a more basic understanding of editing, understood as its central and defining technique—an assumption accepted by Orpen in her study. Orpen does not consider the latest digital editing technology, offering only a brief reference to new "non-linear" systems such as Avid and Lightworks, but it seems to many that an understanding of the digital cinema will equally and necessarily require an understanding of digital editing.[21] Yet, if we are to account for the changes to editing in the digital era, and for their effects on film *as* an art, it will not do to simply declare it aesthetically foundational. While undeniably important, editing is not indispensable, a fact that, as we shall see, has arguably become even more evident in the era of *digital* editing.

Editing is arguably *unique* to the cinema, perhaps even *specific* to it. Berys Gaut, for example, has made such a claim—but he importantly lists several other aspects or techniques that are also unique or specific to film art, none of them, though, presented as either foundational or essential. "Cinema," says Gaut, "is the medium of the moving image; so moving images are unique to it; and certain of its devices are also distinctive, and indeed unique to it. Editing (in the sense of the editing together of motion shots) and camera movement occur only in the cinema, for instance."[22] Gaut is certainly ready to admit that editing is a unique and especially powerful technique, but importantly insists that it is not an indispensable or *essential* aspect of film art. It is a particularly significant formal component of film as an art form, similar, though, and no more or less important than the arguably more basic fact of motion itself, and no more or less significant than an equally unique and alternative technique like camera movement.[23] Editing, as Gaut notes,

20 Ibid., 119.

21 Orpen says that, on a practical and visual level, the principal features of film editing have not changed significantly since its 'invention' at the turn of the century. There has been technological progress," she says, "for example the Moviola in 1924, then the KEM or Steenbeck in the 1960s and now the non-linear systems (AVID, Lightworks)." Ibid., 4.

22 Berys Gaut, *A Philosophy of Cinematic Art* (Cambridge: Cambridge University Press, 2010), 302.

23 I should note that Gaut uses the term "cinema" in a quite expansive sense. By defining the cinema as "the medium of the moving image," Gaut means it to be understood as broadly as possible. I would add that, and even though he describes the cinema as the "medium of the moving image," his account of cinema might seem to exclude but would of course allow for a film like *La Jetée* (Chris Marker, 1962), which in fact consists entirely (with one momentary

offers filmmakers a broad range of formal and expressive possibilities, as a particular "mode" of presentation. The decision to edit a scene, that is, leads to further and more specific artistic choices, the results of which are, we may say, distinctively cinematic, as Gaut notes. "A scene can be edited," he says, "in different ways—the editing mode of presentation of a scene can be varied—so that artistic properties can be achieved that are distinctive to the medium. For instance, by accelerating the editing rhythm one can generate increasing emotional tension and excitement."[24] While "distinctive," though, this does not mean that editing is necessary or essential, or that it should be understood as the singular or most suitable technique capable of creating such cinematic effects, given that other choices, with comparable effects, generating a range of other subsequent choices, are also available, with equally distinctive results, as Gaut observes. He notes that a scene can be constructed with editing, *or*, as he says, "a scene can be shot with different camera movements (or none), so varying the mode of presentation in a way distinctive to the medium. For instance, camera movements can be used to pick out one part of a scene as important; or movements can be created to suggest a personality or a presence; or the camera may be held stationary for long periods so as to suggest a sense of being entrapped."[25]

The point, of course, is that a film is no more or less cinematic depending on which techniques are used, that it is still a work of film art even if it entirely eschews editing, using only camera movements; or even if it eschews both! The design, for example, of Andy Warhol's *Empire* (1963), a single, unedited, eight-hour static shot of the Empire State Building at night, reveals how a film (which I will insist this is) may achieve a distinctively cinematic effect by merely keeping a camera stationary and by not cutting at all. Such total eschewal is rare, though. As Gaut says, without claiming any hierarchical relation between the two, without claiming one or the other to be foundational or essential, "in general visual styles in cinema are partly constituted by particular choices about editing and camera movement."[26] One may, of course, as Gaut implies, and as Warhol did, choose neither to edit nor to move the camera. Such a choice, though, makes artistic sense, we could say, insofar as editing and camera movement are noticeable (and notable) by their absence, given their ubiquity in the history of film. Most

exception) of only still images, and is indeed described in the opening titles as a "*un photo-roman*," a photo-novel or photo-story. There is arguably still *movement*, though, from photo to photo, achieved precisely through editing.

24 Gaut, *Philosophy of Cinematic Art*, 302.

25 Ibid., 302–3.

26 Ibid., 303.

films (the vast majority) use either editing or camera movement, and typically both in some combination, making Warhol's example an obviously anomalous one (not surprising, given that it is Warhol). While literally exceptional, though, the film is no less cinematic for its choice to neither edit nor move the camera, for its explicit eschewal of the most distinctively or historically familiar cinematic techniques. As distinctive techniques, unique to the cinema, specific to the medium, editing and camera movement nevertheless establish, as one artistic possibility among many, the choice *not* to use either of them.

This, as we shall see later, is particularly important to note given that one of the (perhaps ironic) consequences of the advent of digital editing is that one need not any longer edit at all. It has become possible (or, more precisely, much easier) to create ever longer shots or even whole films consisting of a single shot. Indeed, this has become so common, so familiar an aesthetic tendency, that is has been made the very subject of a recent film, *Coupez!* (2022), by the French filmmaker Michel Hazanavicius.[27] It is a self-conscious film about filmmaking, designed with the familiar structure of a film within a film, telling the story of the making of a low-budget zombie movie. It opens with a long, single shot, of about thirty-seven minutes, which is exceedingly poorly executed—but which is revealed to be the risky artistic choice of the fictional filmmaker. He is trying to outdo other filmmakers who have recently been constructing ever longer single shots, often with complex movements and elaborate mobile stagings—a formal cinematic gesture that has increasingly become a mark of a certain kind of artistic bravery (or bravado), a deliberate and conscious *refusal* to edit, even when editing has become arguably much easier and more efficient as a craft with new digital software and systems. The rise in the number of long takes—of long, single-shot, moving camera sequences—a tendency that Hazanavicius is satirizing in his film, is, as I shall discuss later, arguably one of the most obvious stylistic effects of the new digital technologies, which include both digital cameras that are not limited by the amount of film that fits in a magazine, and digital editing systems, which can be used either to process and potentially add other kinds of effects to the long takes, or to make several takes appear to be one long take, a single shot, by effectively obscuring the cuts, rendering them actually invisible. The fact of editing, and of editing in the digital age, is a central theme of Hazanavicius' film. The title, *Coupez!*, may be literally translated as "cut!," in the imperative form

27 I would like to thank Martin Lefebvre for bringing this film to my attention, in one of our many conversations on the matter of technology and technique, the technical and the stylistic.

of the verb, and brings to mind the command made by the film director, to cease shooting at the completion of a scene, but also to the acts that will follow, during post-production, when the film is edited, when it is *cut*.[28] There is also a comic suggestion in the title, though, which is that the fictional filmmaker, having fatefully decided to make the very long opening sequence of his zombie movie a single, uninterrupted shot, should, perhaps, and more wisely, have "cut!" Yet the choice of camera movement rather than editing—to only use camera movement and to entirely eschew editing—is a perfectly legitimate one. The question, ultimately, is only whether one has the artistic ability to follow through successfully with whatever choice one makes as a filmmaker. The importance, but not the indispensability, of cutting is emphasized, in this knowing comic film, as we are presented with the dire results of a choice to forego editing, a refusal to cut—which is presented, though, as a legitimate artistic choice, even if the success of the choice is not guaranteed, even if the results are so comically bad, given the difficulty, the challenge, that *any* artistic choice imposes.

Even if the choice of our fictional filmmaker had been, in the simple infinitive form, *couper*, "to cut," if, that is, the choice had been to construct the opening sequence with editing rather than through camera movement, the artistic challenge would have been just as great, the risks just as considerable. In this sense, the title of the film is a kind of challenge, a dare—*coupez!*, "cut if you can!" It is certainly among the most fateful decisions made by a filmmaker, leading to the many further decisions. That is, having chosen to edit, to construct a sequence with editing, the questions are then when and how to cut, and even when *not* to cut. With the transition to digital editing, the technological context within which such artistic decisions are made now has indeed changed dramatically, with a consequent increase of the range of choices that follow from those initial decisions. The transition is almost complete. Hardly any filmmaker today actually "cuts" film. Editing no longer involves scissors and glue, blades and sticky tape. Directors no longer huddle with their editors over one of the original dedicated devices for editing, the Moviola or the Steenbeck, winding film on spools, peering through a viewfinder, holding the very material of the film itself in their hands, perhaps, and, in this respect, feeling

28 The film has been released in English with the title *Final Cut*, with the loss of such associations, but this is, and perhaps not coincidentally, also the name of one of the most widely used new digital editing software systems. I have not been able to determine, though, if *Coupez!—Final Cut*—was indeed cut on "Final Cut." The film is a remake of *One Cut of the Dead* (Shin'ichirō Ueda, 2017), which is also about a film crew making a budget zombie film, opening with a long, complex single shot, which is also botched.

the fatefulness of the decision to cut, or not, even more acutely. With the latest technological changes, the nature of film editing, as a craft, and as a technique, an artistic decision, has undeniably changed, and considerably so. Indeed, the very names of these earlier editing devices have a distinctly antiquated ring, bringing to mind, for those who remember, whirring gears, spinning reels, the "chunk" of the blade being brought down, the satisfying pressure of the splicing handle. Now, though, virtually all "film" editing is done on a computer, with all of the actions performed by pressing keys on a keyboard or clicking a mouse or trackpad. To cut, "*couper*," is now entirely conceptual, no longer based in the actual physical act of cutting. Of course, it has always been conceptual. A kind of decoupling has taken place, though, so that the technique of editing, of cutting, has been separated, cut off, we could even say, from its original material source. *Coupez!*, the command, as Hazanavicius seems to understand, has a significantly different sense now for editors and filmmakers, with potential consequences for editing style, for editing as a technique. Without a clear understanding of editing *as* an artistic technique, however, it is difficult to make sense of the changes to editing technology or to appreciate the actual consequences or significance of those changes, to describe the aesthetics of digital montage.

***

I approach this, then, precisely as an *aesthetic* matter, considering, that is, the effects that these changes have had on the cinema as an art, on editing as an artistic *technique*. To describe it this way, though, is already to court some complexity, some significant ambiguity, introducing an instance of what Monroe Beardsley once called the "misleading idioms," like "technique," that he argues are so common in the philosophy of art, in philosophical aesthetics and in critical analysis. "The term 'technique,'" he says, "has a habit of skipping lightly back and forth from the aesthetic object to its physical basis."[29] There is something to Beardsley's observation. This becomes clear when we consider what it is to talk about editing in the cinema. Whether in its traditional form or in its most recent configuration, as either the simple act of cutting strips of film into smaller pieces and gluing or taping them to other pieces, or as the now more technically complex process of arranging shots and of potentially modifying the content of those shots with software on a computer, editing as a *technique* is distinct from the experience of an

29 Monroe Beardsley, *Aesthetics: Problems in the Philosophy of Criticism*, 2nd ed. (Indianapolis: Hackett, 1981), 81.

edited film, distinct, that is, from the film as an aesthetic object, derived in some basically contingent way from some physical basis, the operations of which are either invisible to or of no specific interest for the viewer.[30] That a filmmaker has arranged a sequence of shots in a particular order, through whatever means, either with a simple blade, with a pair of scissors, or with a more elaborate device such as a Moviola or on a flatbed, or now with a complex digital computing system, is a fact distinct from and possibly irrelevant to the perceptual and conceptual experience, that is to say, the *aesthetic* experience, of seeing these shots in this particular order. If it *is* relevant, it is difficult to explain how, in what specific respect.

Indeed, it is common to observe that, despite the considerable changes to the technology of editing, little if anything about the cinematic experience, or the specific experience of editing as a technique, as an aesthetic experience, has changed. This is what is initially so puzzling about the transition to digital montage, the fact that, while the *technical* changes to film editing in the digital era have been so significant, while the materials and tools used by the film editor have been so comprehensively transformed, little about editing as an *aesthetic* fact seems to have changed at all. Meraj Dhir, having considered the specific effects of the new digital editing systems, concludes that it is "unlikely that the underlying principles informing classical continuity editing in mainstream cinema [will] change in any radical or fundamental way."[31] He notes that, at the point that he was writing, any changes that had occurred had been slight, and had not changed the fundamental logic of the most common editing patterns or styles. Sergej Ivasović, offering one of the most detailed and comprehensive accounts of the new digital editing technology, similarly concludes that, despite the breadth and significance of the changes to the technical apparatus of editing, the aesthetic consequences are minimal—perhaps entirely negligible. No matter what changes about the tools of editing, he argues, "the main *editing transitions* will remain the same. A cut will still be a cut, pure and simple.

30 This is what is implied in the essay, co-written with Martin Lefebvre, in which I first considered the question of digital editing. "'A cut ... is a cut, is a cut, and so is the impression of a cut.' This is," we said, "how one might characterize the film viewer's reaction to digital editing. For the obvious fact is that cuts 'made' with the help of a digital system look exactly the same as cuts made by hand or with the help of a flatbed." See Martin Lefebvre and Marc Furstenau, "Digital Editing and Montage: The Vanishing Celluloid and Beyond," *Cinémas* 13, nos. 1–2 (Fall 2002): 69–70.

31 Meraj Dhir, "The Modern Entertainment Marketplace, 2000–Present: Editing," in Charlie Keil and Kristen Whissel, eds. *Editing and Special/Visual Effects* (New Brunswick, NJ: Rutgers University Press, 2016), 170.

Comprehensible to everyone. No matter how great and dramatic changes on the level of editing expression," he insists, "editing will remain the same in its essence."[32]

This is despite the fact that it is routinely claimed that digital technologies, and specifically digital editing technologies, have indeed altered the cinema as an art in some fundamental way. "Computer media," Lev Manovich has said, in one of the more famous and influential versions of the claim, "redefine the very identity of cinema."[33] The most significant aspect of this change, the very source of such "redefinition," Manovich insists, is in fact editing, which he proposes as the constitutive and defining technique of the cinema—to which the most significant changes have been applied. He observes that, "if at the beginning of the twentieth century, cinema discovered that it could simulate a single space through temporal montage—a time-based mosaic of different shots—by the end of the century it had arrived at a technique to accomplish a similar result without montage."[34] Traditional editing, that is, has been replaced by what he calls "digital compositing," which he argues is a radical transformation, expanding the power of editing to the shot itself. "While traditional film montage," he says, "privileges temporal montage over montage within a shot ... compositing makes them equal. More precisely, it erases the strict conceptual and technical distinction between the two."[35] The consequence is a radical transformation in our theory of editing, he suggests, and thus in our very understanding of the cinema, a transformation in our understanding of montage as a technique, as it has been conceived, referring to the origins of such theory. "Thus if Pudovkin, one of the theorists and practitioners of the Russian montage movement of the 1920s, conceived of montage as a one-dimensional line of bricks, now it becomes a 2-D brick wall."[36]

Whether Pudovkin in fact thought of editing in such simplistic terms is an open question, and the distinction that Manovich makes here, I will suggest, perhaps needs to be qualified. That both things seem to be true at the same time, though—that little if anything about our experience of the cinema, and of the specific technique of editing in the cinema, has changed, while the cinema as a technical apparatus, as a technological complex, seems so

32 Sergej Ivasović, "Digital Editing," *Moveast*, 7 (2002): 156–57; emphasis in original. The essay was first published in Croatian, in 2001, as "Digitalna montaža," in Vol. 26 of the journal *Hrvatski filmski ljetopis*.

33 Lev Manovich, *The Language of New Media* (Cambridge, MA: MIT Press, 2001), 293.

34 Ibid., 155.

35 Ibid.

36 Ibid., 156.

obviously to have been so comprehensively transformed, including perhaps most significantly the technical apparatus for editing—seems difficult to reconcile. In the face of such difficulty, it is often simply declared that the cinema has been transformed, as a result of the changes to editing, merely on the assumption that editing is the basic source of the cinema's artistic or aesthetic identity. That identity seemed to have depended, as Manovich says, on some strict distinction between the arrangement of shots, on the one hand, and the modification of elements of the shot, on the other. Digital editing seems to confound this distinction. Lucy Fischer had already made this claim, considering the effects of the new digital editing systems that were becoming quite prominent by the late 1990s. The advent of what she calls "new, high-tech methods" have, she argues, precipitated a "crisis in editing"—comparable, she says, to the crisis created by the coming of sound in the 1920s and 1930s. She is quite specific about the change that has taken place. In the past, she says, "film editing meant plastic relationships *between* shots." Now, though, "editing equally implies synthetic relations *within* shots." The consequence, she says, is a "transformation of cinema's traditional association with realism," on which specific grounds she suggests that "a new philosophy or ontology of the medium may be in order."[37]

This view is the result, though, as I have been suggesting, of some unresolved conceptual difficulties in film theory, of the ways in which we think about the cinema as an art form, about films as *works* of art, which are in fact characterized by a formal, aesthetic complexity, which films must derive, in some manner, from the complexity of the technical apparatus for the creation of works of film art. This basic fact about technology and technique has not changed, at least not to the extent that we need a "new ontology," on the assumption that the thing being considered, the cinema, the apparatus of filmmaking, has been transformed in such a way and to such a degree, from a form that some earlier philosophy had been able to understand, that necessitates a new one, presumably still to be imagined or invented.[38] The complexity of the digital cinema is, though, I would insist, in its essential form, the same as what had characterized the so-called photographic cinema, both of which should be susceptible to explanation

37 Lucy Fischer, "Film Editing," in *A Companion to Film Theory*, eds. Toby Miller and Robert Stam (Malden, MA and Oxford: Blackwell, 1999), 81; emphases in original.
38 The question is muddied further by Fischer's off-hand reference to "realism," a term so familiar but conceptually vague. I have considered the concept of realism in specific relation to the technologies of special effects. See Marc Furstenau, "Realism, Illusion, and Special Effects in the Cinema," in Martin Lefebvre and Marc Furstenau, eds., *Special Effects on the Screen: Faking the View from Méliès to Motion Capture* (Amsterdam: Amsterdam University Press, 2022), 131–63.

by a more general ontology of art, according to an account of the nature of the work of art that is as applicable to the cinema of the past as it is to the contemporary, digital cinema, as applicable to the work of film art as to any other kind of work of art. It is with just such complexity in mind that a description of the recent changes to editing technology, the transformation from film editing to digital montage, the shift from the essentially mechanical procedures of cutting filmstrips to the complex processes of computer editing systems, should be undertaken. To the extent that I do so here it is with the hope that such a description may clarify some basic problems in film theory. The most basic, arguably, is indeed the problem of explaining the relation between technology and technique, between, that is, the physical, material, and technical basis of the cinema and the artistic techniques of filmmakers, their film style. Put in more general terms, this is the distinction between *technique*, understood as the means through which material effects that can be created by an artist, to create a style, through specific technological or technical provisions, which are varied and historically variable, and *aesthetics*, the quality of a work of art as it thus experienced, the effects of our encounters with the works of artists created through such means. The challenge, of course, is to clearly explain what the relation is between the technical origins of a work of art and our experience of it as an aesthetic object, difficulties that have beset much film theory, and theories of art more generally.

This is indeed the challenge, the problem, that Monroe Beardsley points to, in order, though, as he so famously did, to distinguish what is legitimate in the critical analysis of art from what is illegitimate. He is concerned about the concept of "technique," and the effect of its use, what he describes as the "skipping back and forth" between two distinct things, between, on the one hand what he describes as the "physical basis" of art, and, on the other, what he calls the "aesthetic object." Beardsley is most famous for his account of what he (with W. K. Wimsatt) called the "intentional fallacy," to be avoided, they argued, so famously, in any critical discourse on art, as a principle of a philosophical aesthetics.[39] This, though, is a part of a broader and more general category that he proposes, the "genetic," which is in fact one of three categories of justification for aesthetic evaluation, which also include what he calls the "affective" and the "objective." The intentional fallacy is only a specific instance of the broader "genetic fallacy." As a kind of reason for judgment, the "genetic," Beardsley explains, "refers to something

39 See W. K. Wimsatt and M. C. Beardsley, "The Intentional Fallacy," *The Sewanee Review* 54, no. 3 (July–September 1946): 468–88.

existing before the work itself, to the manner in which it was produced, or its connection with antecedent objects and psychological states."[40] The problem with the concept of "technique," in this respect, and as Gérard Genette says, is that it tends, usually in some sort of unsystematic way, to reveal little more than an unargued preference for some extraneous aspect of a work, whose origin or genesis is traced back to some contingent fact about its production or some irretrievable psychological fact about the artist's intention. From considerations of production, Beardsley moves to what we can call reception, where the same kind of error is also made, he argues. The "affective" reasons often given to explain the effects of a work of art, Beardsley argues, result in a similar "fallacy," insofar as the critic is offering no more than a reflection on their own feelings about a work, "judging the work by its psychological effects, or probable psychological effects, upon the critic himself, or others."[41] What Beardsley offers as the only valid source of reasons for aesthetic judgment are what he calls (perhaps not surprisingly) "objective" reasons. A reason is objective, he says, "if it refers to some characteristic—that is, some quality or internal relation, or set of qualities and relations—within the work itself, or to some meaning-relation between the work and the world."[42]

The sort of formalism that is manifested in Beardsley's aesthetics, which was indeed the basis for the formalist school of literary criticism, to which he was a significant contributor, namely the so-called New Criticism, is "objective" insofar as it explicitly and deliberately excludes all questions about the "physical basis" of works of art, as well as any feelings we may ourselves have about the work, but also, and more stridently, any consideration of their intentional origins. Attention is to be paid only to the "aesthetic object" itself, the finished work, which is thought to exist, at least as an object of valid aesthetic criticism, independently of any "genetic" or "affective" aspects, as an object of a purely "objective" critical interest. A similar formalism, I will suggest, can be found in much film theory, certainly insofar as there is a debate, largely around the figure of the so-called "auteur," about the significance of intention in the making of works of film art. In much structuralist or semiotic, and psychoanalytic, film theory, the intentional origins or genesis of the work is set aside in favor, arguably, of "affective" reasons, but also, it could be said, in the hopes of establishing something like an "objective" criticism, a formal (or

40 Beardsley, *Aesthetics*, 457.

41 Ibid., 461.

42 Ibid., 462.

formalist) theory of the cinema, and of the works of cinema, as independent aesthetic objects.

These matters are complex ones, and I am not able to address them in any more thoroughgoing way here, except to say that the question of the genetic origins of the cinema, either in some specific "physical basis," or in the intentions of a filmmaker, the psychological inscrutability of which is only compounded by the inherently collective and collaborative nature of filmmaking, are some of the most substantial, and unresolved issues, in film theory, as in art theory more generally.[43] It is such issues, I am arguing, that need to be sorted, though, in at least some provisional fashion, in order to account for the advent of an aesthetics of digital montage, understood as the result of a transformation of the technical apparatus of film editing. The sorts of distinctions that have been established so famously by Beardsley, and, as we shall see in a moment, are operative in much film theory, have become especially acute again as the physical basis of the cinema has indeed been altered, and as those alterations have supposedly increased the intentional powers that a filmmaker is supposed to now have. The physical basis of the cinema, in the past, when it was "photographic," is conceived of as just the kind of material that limited or importantly constrained artistic intention, in what could be described as a kind of inherent manifestation of the intentional fallacy, forcing critics and theorists to attend instead to the objective, material effects of the photographic image itself, or to the cinematic "apparatus." In either case, though, as Wollen argued, a specific kind of "ontology" is posited, according to a kind of confusion that the very concept of "technique" tends to create, as Beardsley contends.

In many respects, the history of film theory can be analyzed as a series of attempts to clarify, in a manner comparable to Beardsley's efforts, the appropriate reasons for proper or informative critical judgment, which has, though, as with Beardsley's formalist approach, led to certain basic confusions about the significance of both the physical basis of art works and the intentions of the artists creating them. As André Bazin so famously declared, with the photographic image, the supposedly primary physical basis of the cinema, "an image of the world is formed automatically, without the creative intervention of man."[44] This is presented as an epochal event, a novel event in the history of art. "All the arts," he says, "are based on the

43 For a good survey of the problem, with a specific consideration of the very complex issue of intention, see Berys Gaut and Paisley Livingston, eds., *The Creation of Art: New Essays in Philosophical Aesthetics* Cambridge: Cambridge University Press, 2003).

44 Bazin, "The Ontology of the Photographic Image," in *What Is Cinema?*, trans. Hugh Gray, 13.

presence of man, only photography derives an advantage from his absence."[45] Bazin, and the other so-called "realist" theorists like him, as Wollen notes, were rejected by later theorists, who argued against the claims made about the "ontological" specificity of the photographic image, specifically any claim about its supposedly more intimate relation to "reality," the supposedly "automatic" formation of images of the world, offering instead an account of the operations of the "apparatus," building on the more "formalist" accounts that had been provided by earlier theorists like Rudolf Arnheim, Béla Balázs, and, significantly, Sergei Eisenstein. Such an apparatus would generate specific effects upon the film viewer, but in a manner, as Wollen also says, that are ultimately, and similarly, conceived as "ontological," insofar as the apparatus was understood itself to be a kind of autonomous force, or the source of autonomously generated effects, the effects of the "system" itself, the formal apparatus itself, operating according to the internal, inherent operational logic of the mechanisms.

More recently, though, there has been a return to Bazin and the realists, whose accounts of the cinema as an essentially photographic medium are the basis upon which the changes with digitization have been assessed, with the sense primarily, as I say, that it has meant an increased intentional power in what had otherwise been understood as a form resistant to intentional control, constrained by the automatic nature of photographic inscription. In both cases, though, an autonomy is posited, either of the image or the apparatus, against which some form of intention, either properly constrained or dangerously unfettered, is arrayed. On the basis of the kind of analysis provided by Beardsley, we can see how film theory has long struggled with the various genetic and affective aspects of the cinema, judging them to be either more or less appropriate in the pursuit of what could be described as an objective aesthetics of the cinema, of film art. The degree to which these issues are unresolved is manifested in the kinds of accounts that have been offered of the digital cinema, and specifically of digital editing, which see it either as a triumph of intentional or formal control, specifically through an enlarged power of editing or montage, or as an unfortunate imposition of an overweening intentionality, as the triumph of editing, as the clearest manifestation of artistic intention, over the supposedly unique nature of photography, thought to constrain intention.

This was the result of what V. F. Perkins described as the "orthodoxy" of so much film theory, which reduced the cinema to its physical basis, to its material origins, which "defines the medium as 'film,' meaning the

45 Ibid.

stuff that goes through the camera, whereas the subject of criticism is actually the movie, the thing we see on the screen."[46] Perkins, as we have seen, argued that such orthodoxy was manifested in the two competing "mystiques" of film theory, that of the "image," and that of "montage." Like Beardsley, Perkins in fact advocates for a proscription against any kind of genetic account of the cinema, arguing that the only thing with which we can legitimately be concerned is, as he says, "the thing seen," the movie, which he effectively posits as the autonomous aesthetic object, comparable to the "text" of New Criticism. The tendency that he was criticizing, though, was one that saw a specific value in some particular aspect of the technical apparatus, whether the means for the production of the photographic images or for their alteration and manipulation through montage. He was correct to argue that either of the orthodoxies resulted inevitably in a normative, prescriptive kind of theory, according to which the filmmaker was supposed to be true to one or the other aspect, to fulfill the inherent nature of the cinema by subordinating any creative efforts to either the demands of the image or to the logic of montage. Such orthodoxies are still more or less in place today, in accounts of the new digital technologies that are thought to have transformed the cinema and according to which there are supposed to be more or less appropriate responses by filmmakers. The answer is not, though, as Perkins seems to suggest, like Beardsley, to set genetic or affective questions aside, in the hope of achieving some objective aesthetic basis for a legitimate criticism, but rather to explain more effectively the actual relations that exist between the physical basis of an art form and the aesthetic objects that are created with it. The tendency in most recent accounts of the digital cinema, though, is to say that, with the new physical basis, a new kind of cinema has emerged. If one had subscribed to the orthodoxy of the image, this is understood in terms of loss, as an imposition of an overweening intentionalism that had been checked by the automatic quality of photographic recording. If one had subscribed to the orthodoxy of montage, it is understood as the final triumph of intention, as the realization of the aesthetic ambitions of the cinema, which had in fact been limited by the constraints of photography. In either case, though, there is an unwarranted ascription of aesthetic value to a particular physical basis, which is thought to produce different aesthetic effects. Rather, as I am arguing, regardless of the specific material from which a work is made, the pertinent fact is the working itself—the fact that all works of art have their origins in the working of some kind of material, that they are derived necessarily

46 V. F. Perkins, *Film as Film: Understanding and Judging Movies* (London: Penguin, 1972), 17.

from some kind of physical basis, the results of which process have less to do with any putative quality that material or basis possesses, but from the fact of artistry manifested in the process.

***

It is in this respect that I am seeking to describe the aesthetics of digital montage, which I conceive of as the effects produced by the working of the material now offered by the cinematic apparatus in its newly configured form, in the kinds of works that are being made and presented just as works of film art, to be judged on just those grounds. Gérard Genette has argued that aesthetics consists primarily in the fact of works being offered for just such consideration, as objects *made*, through specific *technical* means, for consideration *as* works of art. In fact, rather than an *aesthetics* in the classical sense, Genette considers what he describes as the "aesthetic relation," or the "artistic function," of works of art (only one, if particularly important, function of works of art, which can, of course, as he says, have other functions—"practical" functions, whether commercial, political, autobiographical, or what have you). Moreover, the aesthetic is not exclusive to works of art, yet the relation is different depending on whether what we are experiencing is judged to be a work of art or not. While many other objects, both natural and human-made, can, as he says, "trigger" an aesthetic response, works of art are distinct insofar as they "solicit" such a response, insofar as they are *offered*, as he says, as "candidates" for appreciation, conceiving of the aesthetic relation thereby just as deliberations about or evaluations of such "candidacy." On such grounds, Genette argues, "the specific and, therefore, defining feature of works of art is, as I see it, that they proceed from an aesthetic *intention*, and accordingly perform an aesthetic function, whereas other sorts of things can at best elicit a purely attentional aesthetic *effect*."[47] This is perhaps the most significant aspect of Genette's argument, which is clearly distinguished from the Kantian tradition in aesthetics, insofar as Kant refused to countenance the possibility of natural phenomena having aesthetic value (relegated instead to the realm of the sublime). By contrast, Genette is willing to accept that we can very easily find aesthetic value in natural phenomena, to find pleasing to our aesthetic sense a natural landscape, a rainbow, an animal's gesture. Such value is, though, Genette says, "attentional," the result of our *attending* to such

47 Gérard Genette, *The Aesthetic Relation*, trans. G. M. Goshgarian (Ithaca: Cornell University Press, 1999), 1; emphases in original.

objects or phenomena in such a manner, as a choice that we can make. By contrast, what distinguishes works of art is that the relation is "intentional," in that they are deliberately, intentionally offered as "candidates" for aesthetic appreciation, deliberately made *as* such offerings. "What is uniquely characteristic of works of art," Genette says, "is their intentional aesthetic function, that is, their artistic function."[48]

This can, he says, be put in more "subjective terms," such that "what endows an object with the status of an artwork in the eyes of its receiver is the impression, warranted or not, that it proceeds from an intention that was at least in part aesthetic."[49] I will simply accept this, at this point, as axiomatic. I will not be offering a thoroughgoing defense of such a claim, but will, as I proceed, do so on the basis of at least this basic assumption—that, like any other artist, the filmmaker, the film artist, undertakes to create a work of film art intentionally, to offer it as a "candidate" for aesthetic appreciation and judgment. Moreover, the filmmaker does so through some specific technical means—but this, and as Genette argues, is no different than any other artist, working in whatever other medium, with any kind of material or physical means, with whatever tools or technology. "An artwork," Genette says, developing his specifically intentional aesthetics, "is an object that is simultaneously (and intentionally) aesthetic and technical; its technical properties are aesthetically relevant to the extent that they shape our appreciation of this work."[50] This, in fact, is a central aspect of Genette's argument, which I will also accept, as a guide in my consideration of the technical qualities of editing as a part of the apparatus employed by film artists to create works intentionally to be offered as candidates for aesthetic appreciation. Genette expressly emphasizes the relation between aesthetics and technique, between the technical and the artistic. Works of art, he argues, on the one hand, "by virtue of their specific intentionality (their 'candidacy') ... solicit, exclusively or over and above whatever practical function they may have, a more constitutive aesthetic attention than that which may arbitrarily be brought to bear on purely attentional 'aesthetic objects' (natural objects, utilitarian artifacts)."[51] On the other hand, as he notes, "by virtue of this very intentionality, they mobilize technical means in its service, which, if they come to our attention, have an incidence on the work's aesthetic function, when, and insofar as, they influence our

48 Ibid., 2.

49 Ibid.

50 Ibid., 151.

51 Ibid., 154.

appreciation of them."[52] Genette's argument in fact comes down to this, to an emphasis on the technical and material aspects of art *just insofar* as they are the very means through which intention is realized, in the process of generating artistic candidacy. "Indeed, it seems to me," he says, "that this aesthetic relevance of technical data constitutes the essential feature of the special status of artworks."[53] Genette provides a concise summary of his view, which is the basis upon which I undertake my analysis of the aesthetics of digital montage:

> The artistic function is, par excellence, the locus of interaction between the aesthetic and the technical. The same thing can, of course, be said the other way around. *There exists* an artistic function *whenever* the technical and the aesthetic are conjoined: whenever an activity, and therefore a technical factor, produce an aesthetic effect by influencing aesthetic appreciation.[54]

This has long been acknowledged as a problem in film theory, which has sought to explain the relation between the technical and the aesthetic, but with the tendency to *reduce* the aesthetic to the technical, or to derive aesthetic value from some particular aspect of the technical apparatus, which would, however, be elevated over any other aspect and regardless of the fact of intention. This is, I would submit, what has made it so difficult for film theorists to contend with technological innovation, which so often seems to threaten some privileged technical aspect of the apparatus. Well before the advent of digital technologies, Stephen Heath was concerned with the problem of technological innovation in the cinema, with the confounding nature of the relation between the technical and the aesthetic, between the cinema as a technological apparatus and as an art form, and he presents the problem in such a way that, somewhat surprisingly perhaps, recalls the terms of Beardsley's argument. "Within cinema," Heath wrote, "*techniques* are procedures involving elements of the technology in specific ways, processes in the production and presentation of films."[55] Yet this is not nearly as straightforward as it seems. "The terminology is wavering," he admits, "confusing, since not all of the techniques involved in the

52 Ibid., 155.

53 Ibid.

54 Ibid., 155–56; emphasis in original.

55 Stephen Heath, "The Cinematic Apparatus: Technology as Historical and Cultural Form," in *The Cinematic Apparatus*, eds. Teresa De Lauretis and Stephen Heath (London: Macmillan, 1980), 10.

production-presentation of films are dependent on technological processes for their operation."[56] As an obvious example, he says, the "techniques of acting play a part in the production of many films without necessarily being dependent on the technology of the cinema."[57] Even more confusing are effects that would seem to depend directly on technology but which, upon analysis, are shown to be independent of it. This is particularly significant when considering the "technique" of editing, or, more precisely, specific editing techniques, like the jump cut, which, Heath says, "does not depend on a technological process in the manner that, say, a zoom shot depends on a certain type of lens."[58] He presents this as a general problem of film theory:

> Within cinema, as it were, the question of the relation between technological determinants (the inventions, advances, improvements, modifications) and techniques (the standard or individual practices) in which technology is exploited is a question at the level of films: the determination of the forms of films by technical elements or/and the determination of the technical elements by the aesthetic requirements of film form, "technique" being the term that shifts between the one and the other.[59]

Beardsley had made more or less the same observation, when he said that the concept of "technique" has the tendency of "skipping lightly back and forth from the aesthetic object to its physical basis." This suggests, as I say, a perhaps surprising coincidence of thinking between the radical film theorist and the more traditional aesthetician. For Heath this is supposed to be the result of the inherently technological nature of the cinema, but Beardsley insists that this is a problem characteristic of art in general, a problem of critical terminology when discussing any art. It is not any more acute, as Heath and other such "apparatus theorists" in fact suggest, in a supposedly more technological art like the cinema. "In the fine arts," says Beardsley, "technique" as a term "often refers to such things as the movements of the painter's hand, his method of applying pigment; in this sense, there is no such thing as a composer's technique, though there is a performer's technique."[60] In another sense, "technique" refers to specific sensory or perceptual aspects of a work of art, for example, to "things that can be

56 Ibid.
57 Ibid.
58 Ibid., 10–11.
59 Ibid., 11.
60 Beardsley, *Aesthetics*, 81.

heard in the music itself, to certain kinds of dissonances or the relations between voices in counterpoint."[61] Similarly, it can refer to "what can be seen in a painting—say, size, direction, or width of brush strokes, or thickness of paint."[62] When using the term in any of these senses, to identify any of these aspects of a particular work of art, a critic may, he says, make it clear from the context which kind of technique is being considered, or whether it indeed qualifies in the strong sense of the word as a "technique" or not. Yet, even so, there is an inherent ambiguity embodied in the word itself. "The main objection to the term," Beardsley says, "is that its use claims that there exists a distinction that has, in fact, never been made. Suppose you copied out a thousand descriptive statements about paintings and musical compositions; by what criteria could you decide which of them are to be classified as statements about technique instead of about something else?"[63]

The problem is especially acute, I would argue, in much film theory, and in many critical analyses of film, which proceed on the basis, sometimes explicit, more often implicit, of certain theoretical assumptions about certain preferred techniques. As Beardsley suggests, distinctions are implied that have not been made when discussing film, when considering the various so-called "techniques" of the cinema. To take Heath's example, is a jump cut a technique? In what respect is it determined by any particular technology? It is, arguably, unique to the cinema, as a technique, but it is possible to achieve whether one is editing with a pair of scissors or with a digital editing system, neither of which in any obvious way encourages the use of the "technique" more than the other. What, then, is its precise relation to any particular technological aspect of the cinema, or even any specifiable part of the specific technology of editing? In what way could we, for instance, make any predictions about or even offer any generalizations about what kinds of new or modified techniques any technological innovation in the cinema might produce—like the jump cut? How would an explanation of the technique of the jump cut be reduced to, or traced to, any specific technological aspect of the cinema? Indeed, to describe

61 Ibid.

62 Ibid.

63 Ibid., 81–82. Hence Beardsley's contention that we can never take intention into consideration. Yet, as Genette argues, in a lengthy consideration of Beardsley's claims, we need not reduce intention to a psychological matter, seeking empirical manifestations of any particular intention in specific techniques, but see it rather in more general terms, as the basic grounds of aesthetic judgment, understanding works as intentionally offered for consideration, as "candidates" for aesthetic appreciation, as opposed to our "attentional" decision to admire a sunset or a flower or a snowflake. See Genette, *The Aesthetic Relation*, 73-84, for his arguments against Beardsley.

something like the jump cut *as* a technique, to imply, as this term naturally implies, sharing the same etymological root, that there is some specific *technological* determinant that makes it possible (or even encourages its use) as a technique, is to effectively obscure the more difficult problem of explaining the relation between the material basis of an art and the artistic possibilities, the aesthetic identity, of the art. On Heath's account (and he seems to recognize this problem), only a technique like a zoom shot could be considered truly cinematic, insofar as its formal or aesthetic possibility can clearly be traced to (and reduced to) the specific and exclusive technical possibility provided by a particular lens technology. Without the cinema, of course, as a collection of specific technologies, there are neither jump cuts nor zoom shots—they are specific to, or unique to, the cinema. Jump cuts, though, are possible, as an instance of editing as a technique, regardless of any particular technological configuration of editing technology, while zoom shots are possible only if a particular configuration of camera technology exists. Heath, though, and along these very lines, says that the example of the jump cut is in fact "unconvincing," that "jump cuts are after all bound up with the technology of the cinema, apart from which, unlike techniques of acting, they could have no possible realisation—and thereby at the same time indicative of the distinction to be followed in discussion between cinema and film."[64] I am not sure that his argument is convincing in the end, but he is insisting on the perhaps important distinction between the cinema, or film, as an enabling or determining technological structure, and individual "films" as the particular forms that any such determined instances takes—the latter necessarily traceable to the former. Heath is trying to salvage the concept of technique as a way of explaining the specific effects of the cinematic apparatus, in order to understand the relation between the cinema *as* an apparatus and the specific formal possibilities in any individual film, imagining "a textual analysis informed by reflection on technology/technique (where the answer, of course, could not lie in the simple adoption in the analysis of technical terms, odd references to depth of field or whatever)."[65]

It is a tall order. Most accounts of technique in the cinema in fact, and so easily if usually only implicitly, proceed precisely by "simple adoption of technical terms," by simply offering the "odd reference" to this or that technical fact, using some convenient short-hand phrase or term that is, inevitably, offered on some normative grounds, meant as more or less

64 Heath, "Cinematic Apparatus," 11.
65 Ibid.

appropriate, significant, or essential, when considered against any other technique or technical aspect. I don't think it is by accident that Heath, in this critical passage, singles out the specific technical term that he offers by way of example, namely depth of field ("or whatever"!), which must inevitably call to mind one of the most significant film critics whose criticism is based precisely on a deliberate distinguishing of formal techniques, and of a hierarchical arrangement of them, namely, that is, André Bazin. His implicit theory of film is, I would suggest, established upon a normative ranking of techniques, and his "ontology" of the cinema is based on a privileging of one technique (or one kind of technique) over another (or others)—that is, it proceeds as an argument through the "adoption" in his analysis of privileged technical terms, namely those associated with the initial process of photographic recording, emphasized by such "techniques" as depth of field, of shooting in depth. Photographic recording, and the techniques deriving from and emphasizing that putatively basic, "ontological" fact, is privileged over any other aspect, but specifically over editing, and the techniques that derive from it, from the techniques of montage. It is camera technology, specifically (and lens technology, more specifically, we might add), that provides for the possibility of Bazin's preferred technique (or one of his preferred techniques, but certainly the most significant, for Bazin), namely, depth of field, the hallmark of those filmmakers whose work, whose artistic choices, Bazin celebrated, and who he thought contributed to the development of film as an art, to its aesthetic progress. He highlighted several significant examples, but one is especially important for him. "The influence of *Citizen Kane* cannot be overestimated," wrote Bazin, so famously and influentially. "Thanks to the depth of field, whole scenes are covered in one take, the camera remaining motionless."[66]

He found an alternative, that is, to editing (and to camera movement). Bazin's goal in singling out Orson Welles' film, and the specific technique of depth of field, of the shot in depth, is to reveal its role in the "evolution of film language," and specifically the strides taken by those filmmakers, like Welles, but also William Wyler and Jean Renoir, whom Bazin also mentions, who were looking just for alternatives to editing, to montage, significantly in the early years of the sound era, when naturalistic, synchronized sound had been added to the basic repertoire of filmmaking, making the stylistic flourishes of "montage-style" editing more difficult, but thereby revealing the very aesthetic limitations of editing as a technique. For Bazin, editing, or

66 André Bazin, "The Evolution of the Language of Cinema," in *What Is Cinema?*, trans. Hugh Gray, 33.

montage, had assumed too great a prominence in the silent era, the formal extravagances of which could not be easily incorporated into the sound film. As a result of this specifically technical contingency, he argues, given that it had simply become more difficult to construct the elaborately expressive editing patterns, characteristic, for example, of the Soviet filmmakers, but adopted widely in the silent era, filmmakers now sought expressive alternatives that would work with or complement the new aural aspect of cinematic representation. Not just an alternative to the elaborate patterns of expressionistic editing, though, the shot in depth, or depth of field, was, for Bazin, a signal achievement, a demonstrable advance, obviously superior to the distortions of montage when taken to too great an extreme. "Montage as used by Kuleshov, Eisenstein, or Gance," Bazin says, "did not show us the event; it alluded to it."[67] Editing, montage, inevitably led to what Bazin suggests is a kind of formal exaggeration, becoming a blunt instrument, subordinating the cinema to the intentional, expressive desires of the filmmaker, at the cost, he implies, of the interpretive freedom of the viewer, of their affective response. "The meaning is not in the image," he says, describing the ultimate consequences of montage theory, "it is in the shadow of the image projected by montage onto the field of consciousness of the spectator."[68] In this respect, depth of field must be distinguished as no mere technical device, but rather as a *technique* on par with but in fact superior to montage—or at least as a technique that reveals the "limitations of montage."[69] This is a signal instance of the process described by both Beardsley and Heath, by which "technique" effectively becomes an honorific rather than a truly effective term of critical analysis. Bazin explicitly establishes a hierarchy of techniques, admitting that they all have their place, but insisting that some are more clearly a manifestation

67 Ibid., 25.

68 Ibid., 26.

69 This is a phrase used in another essay by Bazin, "The Virtues and Limitations of Montage," in *What Is Cinema?*, trans. Hugh Gray, 41–52. In this essay, Bazin says he is concerned to "make a simple analysis of certain rules of montage as they relate to cinematic expression, and more specifically still, of its aesthetic ontology" (43). He considers as examples the films *Ballon Rouge* (Albert Lamorisse, 1956) and *Une Fée pas comme les autres* (Jean Tourane, 1956). "Both," he says, "are marvelous examples, in exactly opposite ways, of the virtues and limitations of montage" (43). It is from this passage that Gray derives his title for the essay in English, which, in French, is more direct: "*Montage Interdit*," which Timothy Barnard has translated more literally as "Editing Prohibited." Bazin prefers the long shots in *Ballon Rouge* over the editing effects in *Une Fée*, which is exemplary of the kind of montage he argues should be "prohibited," contrary as it is to the "aesthetic ontology" of the cinema as he understands it. See André Bazin, "Editing Prohibited," in *What Is Cinema?*, trans. Timothy Barnard (Montreal: Caboose, 2009), 73–86.

of the true aesthetic nature or "ontology" of the form. "It would obviously be absurd," admits Bazin, "to deny that montage has added considerably to the progress of film language, but this has happened at the cost of other values, no less definitely cinematic."[70] Bazin is clearly distinguishing his favored technique, when compared to editing, but to other techniques as well, arguing that "depth of field is not just a stock in trade of the cameraman like the use of a series of filters or of such-and-such a style of lighting, it is a capital gain in the field of direction—a dialectical step forward in the history of film language."[71]

Bazin, of course, is rejecting the basic premise of the Soviet montage theorists and filmmakers, for whom editing was the "foundation" of film art. For Bazin, and the other so-called "realist" film theorists, the foundation is to be discovered elsewhere, not in the relations between images, as established by editing, but, rather, in the image itself, and specifically in the *photographic* origins of the image. Discussions of digital editing take place within the context of, and according to the basic terms established by, this particular debate in film theory—between, let's say, with risk of simplifying, but in the form that is usually employed, Eisenstein and Bazin, between editing and photography, between montage and the image—which is, I suggest, still unresolved, and which, I will further suggest, has limited our ability to fully account for the kinds of changes that have occurred to the technical apparatus of editing.[72] There are still really only the two basic views on the question of the origins of film art, on its essential foundation, understood either to derive its identity from the deliberate, intentional techniques of editing, of montage, or from the unique, photographic "ontology" of its raw material. Indeed, the general tendency in most discussions of digital editing, as of the so-called "digital cinema" in general, is to pose the problem precisely as *ontological*. For many who have written on the subject, the shift to systems

70 Ibid., 35.

71 Ibid.

72 Bazin famously presented this as the distinction "between two broad and opposing trends: those directors who put their faith in the image and those who put their faith in reality. By 'image' I mean here," he explains, "very broadly speaking everything that the representation on the screen adds to the object there represented." He clarifies the distinction further, which, he says, "can be reduced essentially to two categories: those that relate to the plastics of the image and those that relate to the resources of montage, which, after all, is simply the ordering of images in time." Without accepting the assumptions behind his distinctions, this indeed describes the two main positions in much film theory, both "classical" and "contemporary," which, as I say, distinguish between recording and editing, between image and montage. I will pursue this further below. See André Bazin, "The Evolution of the Language of Cinema," in *What Is Cinema?*, trans. Hugh Gray, 24.

of digital editing is thought to have altered film art in fundamental ways, at the very level of its *ontology*—as that term has been bequeathed to us by Bazin, for whom a relatively straightforward (and determinative) relation is assumed to exist between technology and technique. Most accounts of the digital cinema see a similarly direct relation between material change and conceptual status. Lev Manovich carefully describes the end of the cinema and the rise of the era of "new media," the "language" of which would now determine all related, or now subordinated, and ostensibly "old" media, especially film, arguing that "the differences between cinematic and synthetic realism begin on the level of ontology."[73] D. N. Rodowick describes the new condition of the cinema in the digital era, what he called the "virtual life of film," posing this fundamental question: "As digital processes come more and more to displace analogical ones, what is the potential import for a photographic ontology of film?"[74] He argues that cinema had derived its identity from photography, and that "photography's principal powers are those of analogy and indexicality."[75] Digital imagery is different. "Because the digital arts are without substance," he says, "and therefore not easily identified as objects, no medium-specific ontology can fix them."[76] Thomas Elsaesser considers the lingering influence of the cinema in the digital era, the forms and effects of which are now "simulated" by new computer media, but argues that new digital media create a "rupture," radically altering if not destroying the cinema's identity, its ontology, even as the cinema may seem to be unchanged, remaining in what seems, at least perceptually, to be a similar form but which, through the digital process, is only *simulating* the basic effects of the cinematic apparatus, without maintaining the photographic integrity of the apparatus. "The digital as rupture," says Elsaesser, "thus leaves us with a paradox: just as it installs the cinema as episteme or ontology, it does away with cinema as a unique technology of imaging."[77]

There does indeed seem to be something very different about the digital cinema, and perhaps especially about editing in the digital era, which is where the challenge to the unique form of "imaging," thought

73 Manovich, *Language of New Media*, 196.

74 D. N. Rodowick, *The Virtual Life of Film* (Cambridge, MA: Harvard University Press, 2007), 9.

75 Ibid.

76 Ibid., 10.

77 Thomas Elsaesser, "Afterword–Digital Cinema and the Apparatus: Archaeologies, Epistemologies, Ontologies," in *Cinema and Technology: Cultures, Theories, Practices*, eds. Bruce Bennett, Marc Furstenau, and Adrian Mackenzie (Basingstoke and New York: Palgrave Macmillan, 2008), 239.

to be characteristic of the cinema, is supposed to be most acute. It is no longer simply a matter of arranging shots in some particular order, but also, it seems, of arranging elements within the shot itself. Manovich has famously and influentially argued that, with the new powers provided by the new media technologies, the fact of editing is transformed most significantly, that "digital filmmakers can create what I will call *spatial montage*"[78] Rodowick has developed the concept further, explicitly building on Manovich's concept (but with more misgivings), offering as alternatives to the traditional concept of editing the notions of "digital compositing" and what he calls the "digital event," which, he says, "corresponds less to the duration and movement of the world than to the control and variation of discrete numerical elements internal to the computer's memory and logical processes."[79] Indeed, he suggests that the most basic element of editing has vanished, arguing that "there may be 'takes' in digital cinema, but there are no shots. Or rather, what was previously considered a shot has now become a highly variable element open to interactive manipulation at the most discrete levels."[80] The shot, and any traditional understanding of editing with the shot as the basic material, an integral and essentially inviolable unit of articulation, is replaced by the "digital event," which is, he says, "any discrete alteration of image or sound data at whatever scale internal to the image."[81] The shift to a digital cinema, for Rodowick, is ultimately manifested, as he puts it, somewhat paradoxically, as I have noted, in the "*devaluation of filmic editing*."[82] If editing was once understood as the arrangement of the integral unit of the shot, thereby properly limited (properly valued) in its capacity for manipulation, it is now an unfettered (devalued) process of alteration and modification that functions at a deeper, lower level, allowing for complete control. "Analog 'shots' define discrete spatial wholes," Rodowick says, "and thus film editing entails combination of these wholes."[83] The breaking down of such "spatial wholes" into smaller units of articulation and expression, he argues, transforms our understanding of editing, and, in turn, of film itself as an aesthetic phenomenon. "Where film editing may be characterized as the addition of spatially discrete wholes (shots; automatism of succession), the modularity and discreteness of the numerical image allow it to be separated into or combined with a

78 Manovich, *Language of New Media*, 158; emphasis in original.
79 Rodowick, *Virtual Life of Film*, 166.
80 Ibid.
81 Ibid., 167.
82 Ibid., 170; emphasis in original.
83 Ibid.

potentially infinite number of digital events that can be manipulated in a great variety of ways."[84]

Rodowick argues that, as a result of these specific technological changes, we are confronted by what he describes as "the curious ontology of digital worlds."[85] He argues that the most significant effects of digital technology are to be found in new editing systems, which create what he says are new process and modes of production, the most basic of which is what he calls "transcoding," that alter the very identity of the cinema, that destroy the basically photographic nature of the cinema, what he describes as its "automatisms." The cinema, he suggests, may appear in many respects to have been unchanged in the process of digitization, noting that "the language of digital creation is predominantly the language of movies. Thus, one way in which film's virtual life extends itself is through its rebirth as the predominant aesthetic template for computational interfaces and works."[86] While superficially similar, though, a more fundamental change is supposed to have taken place. "Cinema," he says, "does not remain unchanged in the process. Through transcoding, filmic automatisms lose their concreteness as creative potentialities and become metaphors for digital functions. Splicers, glue, and editing tables disappear into the screen interfaces of Premiere or Final Cut Pro." Yet this is, of course, a part of the more comprehensive changes, as Rodowick notes. He predicts, in 2007, what was about to come. "The next ten years," he said,

> may witness the almost complete disappearance of celluloid film stock as a recording and exhibition medium. The celluloid strip with its reassuring physical passage of visible images, the noisy and cumbersome cranking of the mechanical film projector or the Steenbeck editing table, the imposing bulk of the film canister are all disappearing one by one into a virtual space, along with the images they so beautifully recorded and presented.[87]

He asks what will be left of the cinema if "it is replaced, part by part, by digitization?" He says that, with this change, "for the first time in the history of film theory the photographic process is challenged as the basis of cinematic representation," noting that the cinema, in what is usually called its analogical mode, "had been defined by a process of the mechanical recording

84 Ibid.
85 Ibid., 171.
86 Ibid., 133.
87 Ibid., 8.

of images through the registration of reflected light on a photosensitive surface."[88]

The problem is again posed as a question about the supposedly "photographic ontology of film."[89] Rodowick's answer is that, and as he tries to show, "a new aesthetic, and a new ontology, of motion pictures are emerging or have already emerged."[90] He argues for what he describes as "the ontological distinctiveness of analogical and digital processes."[91] It is just such a claim, though, that I dispute, based on the assertion that ontology is not derived from materiality. The cinema is characterized, that is, neither by an analogical nor a digital "ontology." Indeed, if we are to use the term at all—a fraught term, rarely argued for or explained, and more often simply deployed—the cinema's "ontology" is an *artistic* ontology. This is enough to contend with, without trying to determine which of the very many different materials of the cinema, as a physical apparatus, are the ones that, if altered, will correspondingly alter the very being, the very identity of the cinema. An ontology of the cinema would have to be grounded upon other assumptions, acknowledging what is in fact common to all of the arts, the shared properties that characterize the arts in general, but also showing in some perspicuous way what is distinctive about art, as a human endeavor, all art, including film art.

I would want to contest the notion of a distinct digital ontology as opposed to another, supposedly different, ontological identity that the cinema is supposed to have possessed in the so-called "analog" era, when the cinema, as Rodowick says, building upon a longstanding tradition, was *photographic*. It is an argument that has become commonplace in film theory, but it is, I want to suggest, a misleading one, the result, primarily, of confusing the distinction between the technical and the aesthetic, between the physical basis of an art and the aesthetic object, between technology and technique. I will put the matter as simply as I can, first of all, and in the most general terms. If we are to discuss film art, if we are to consider the works of the cinema as works of art, then the ontology of the cinema *is* the ontology of the work of art. Works of art share a common ontology, regardless of the specific material used in the making of the work. Even a specific art form is materially heterogeneous, such that instances of that art form, specific works of art in that form, can be made out of any number of different materials

88 Ibid., 9.
89 Ibid.
90 Ibid., 110.
91 Ibid., 113.

or substances. This is obvious in, for example, painting and sculpture, and is no less so in the cinema. Whether films are made photographically, or with hand-drawn images, with a camera or without a camera, with digital technology or with any other kind of technology, whether films are edited mechanically or on a computer, shot on film or video, they remain works of art, works of *film* or *cinematic* art—at least insofar as we continue to use the general category of the "cinema," as the word to describe the various and constantly changing means available to film artists to make specific works of art, what we still tend to call "films."

The "ontology" of the cinema *is* the ontology of the work of art. It is no simple matter, of course, determining just what the ontology of the work of art is. Joseph Margolis has described what he calls "the ontological peculiarity of works of art." Margolis rejects any universal conception of art, of the work of art understood as a "universal," as the manifestation, somehow, of some properties derived from anything other than human intention, or from the basic, physical fact of a human being working with some specific materials. While such basic facts are significant, contributing to the form that any individual art work will assume, none is to be thought of as defining or essential, as the source of the art's being or identity. A work of film art is not a work of art because it is "photographic," that is, nor because it is "digital." The various material sources of cinematic representation are not the source of some unique aesthetic identity. A work of film art is a work of art because somebody (or group of somebodies) made it—either out of the familiar materials, and in the familiar fashion, that others make such works, or out of entirely different stuff, and in very different ways. "A work of art," as Margolis says, "is a particular."[92] It is not, as is often argued in accounts of the ontology of art, a "universal," and this, as he says, very simply, is "because it is created and can be destroyed; and because it possesses physical and perceptual properties."[93] What those physical and perceptual properties are, though, we could say, is *immaterial*—in both the literal and figural sense. That is, *whatever* a work of art is made out of, it has the properties of a work not because of any specific material, any specific physical basis, but rather because somebody used *something* to make it (in the hope, of course, that nobody will then want to destroy it, but we need not say any more on that tragic question ...). A work of art is a particular, insofar as it is made of some particular material, that makes

92 Joseph Margolis, "The Ontological Peculiarity of Works of Art," in *Art and Philosophy* (Brighton: Harvester Press, 1980), 22.
93 Ibid.

it available to us to consider it more generally as a work of art, according to what Genette describes as a gesture of candidacy. No *specific* material, though, provides the work of art, or "Art," with any universal value. This is, of course, I want to say, the basic error made whenever it is claimed that the cinema is "photographic," or is the "art of editing," or that it is now "digital," and was once "analog." The cinema is what we collectively understand it to be, insofar as artists, filmmakers, present us with works of art *in that name*, considering its value as a work of film art in each particular instance, proffered for such consideration.

"A work of art, then," Margolis goes on to say, "is a particular. But it is a peculiar sort of particular because unlike physical bodies, (i) it can (and must) instantiate another particular; and (ii) it can (and must) be embodied in another particular."[94] That is, and to put it simply for the moment, it must assume its identity *through* some physical means or other, but it does not derive its identity *from* any particular physical material. If we are to consider the cinema an art form, and films as works of art, they must, I contend, conform to this account of their "ontology," which is what Margolis describes as an ontology of cultural emergence, with works of art identified not as universals but as "culturally emergent entities," as he puts it, that we consider aesthetically, in whatever particular form they take. Yet this in itself is in fact to say very little, as Margolis acknowledges. "The ontological characteristics assigned here," he says, "are no more than the most generic characteristics of art; its distinctive nature remains unanalyzed."[95] It is the case not only that much film theory stops the analysis at this point, in fact failing typically to qualify the film as a work of art in this particularly salient ontological sense, as a *particular*, but tends not to proceed to the more subtle and complex questions of the distinctive nature of the film as a work of art, drifting, often, into the realm of the *universal*. Much film theory, that is, when describing film, the cinema, declares its nature as, say, "photographic," or now as "digital," but without explaining how its nature is derived from any such particular physical basis, nor how that is pertinent in any distinctively aesthetic way, in particular works, particular instance of film art.

Margolis has in fact considered these matters in relation to the cinema, tracing the history of the basic concepts of film theory, noting that a "lot of mischief lies buried in the elementary blunder of conflating the physical

94 Ibid.

95 Ibid., 22–23.

and 'aesthetic' features of an artwork—even a film."[96] Margolis is considering specifically the arguments of early, "classical" film theorists—Rudolf Arnheim, Siegfried Kracauer, André Bazin, Stanley Cavell, but also Erwin Panofsky and Walter Benjamin. He considers a passage from Benjamin, from the famous essay on "mechanical reproduction," in which Benjamin makes a principled distinction between the cinema and theater, emphasizing the differences, arguing that, in the case of the cinema, the performances of actors "are presented by the camera," and noting that "the camera continually changes its position with respect to the performance," and that acting in the film "comprises certain factors of movement which are in reality those of the camera."[97] Margolis sees here "a bit of ingenuity and some disingenuousness. ... For one thing, the camera is not doing any of the artistic work; surely it's the director and his or her crew, using the technology of film (which is simply different from the technology of the stage), who are composing and creating a motion picture."[98]

It is upon this very simple basis, that motion pictures are created by filmmakers, by people, using the technology of the cinema, in its various forms, as it changes historically, and to a range of essentially unpredictable effects, that I will proceed to make my arguments about the contemporary technology of editing, about the aesthetics of digital montage. That an artist may choose to make works of art with the tools and technology of the cinema is a choice no different, essentially, than the choice to use the various technologies of the theater. (They may, indeed, combine them, too, if they like, as in the case of the dramaturgy of a cinematically minded theatrical director like Robert Lepage.) The implication, in Benjamin's argument, is that the cinema is a material advance, artistically and aesthetically, but also politically, socially, and culturally, on the theater, that the modern technologies of the cinema are superior in some way to those of the theater. The principle upon which I am arguing here, though, is that all the arts are fundamentally the same in this very basic respect, as technological configurations that provide specific affordances to artists, having chosen to work in one or the other, subsequently offering the completed works for aesthetic candidacy, as Genette argues.

96 Joseph Margolis, "Mechanical Reproduction and Cinematic Humanism," in *What, After All, Is a Work of Art?* (University Park, PA: The Pennsylvania State University Press, 1999), 106.

97 Quoted in ibid., 112. See Walter Benjamin, "The Work of Art in the Age of Mechanical Reproduction," in *Illuminations*, ed. Hanna Arendt, trans. Harry Zohn (New York: Schocken Books, 1969), 228.

98 Margolis, "Cinematic Humanism," 112.

If, again, we are to use the term, art is defined "ontologically" by the fact of choosing particular technological affordances in order to accomplish specific artistic goals—to create works of art as artistic particulars, judged, experienced, evaluated, enjoyed (or dismissed, rejected, or even destroyed) as they are offered for aesthetic consideration. Specific arts offer various affordances, and each offers distinctive kinds of experiences, as Margolis acknowledges, considering the particularly effective affordances of the cinema, which can in fact be employed to produce powerful results. One of the most notable, as he notes, is indeed editing. As Margolis says, "the motion picture commands our sensibilities in unprecedented ways," with the capacity specifically, as he says, "to share anonymously with others a montaged world abstracted from every possible cultural source and fused again with each such source—according to our grasp of cultural reality."[99] He admits that "film is an extraordinary medium. It compels our absorption, even when it is less than at its best, and even when we are distracted."[100] This is largely the result of the ability, in the cinema, to create dynamic, complex moving imagery, and then to edit the results together, in a manner, though, that is not fundamentally different from other kinds of artmaking, nor separated from the ways in which we experience the world, reality, and from the ways that we comport ourselves in the world. Still, the editing in film is the source of many very creative possibilities, for the elaboration of particularly effective artistic expressions, through a kind of explicit gesture of combination of often disparate elements. In this respect, Margolis establishes a basic affinity between editing and our own efforts at making sense of diffuse and disparate cultural material. "In the filmic experience," he says, "montage (viewed in a sense akin to Sergei Eisenstein's) habituates us to the near tolerance of the continuity between the 'real' world and its filmic counterparts."[101] He is not, though, going to make the argument that film is *unique* in any of these respects, only that its particular affordances are unique, that film has its own ways of creating such experiences for us, especially with the resources of "montage," which for Margolis is the specific procedure of editing, but also a more comprehensive process (as proposed by Eisenstein) of combination and assemblage, of construction and simulation. The discoveries that are possible (but by no means guaranteed) by such means, in what is undeniably the very popular form of cinematic representation, are not unique to it, he says, noting that, rather than watching

99 Ibid., 118.
100 Ibid., 119.
101 Ibid.

a film, "we could easily make the same discoveries by reading, say, Marx or Nietzsche, or by dwelling on psychopathologies or epochal histories, or by frequenting the theater or reading literature."[102] Margolis argues that we should think of the arts as "the *utterances* of culturally apt selves," adding that "the arts collect certain notably absorbing utterances that become our own evolving montaged sensibilities apt for further such absorption. Film is simply the most obvious, the most accessible, the most transparent metonym—and engine—of that process in a mass society."[103]

It is with such an account of art, and of film art—without, of course, necessarily accepting all of the consequences of Margolis' more general arguments—that the specific technical development in the cinema, the changes to the tools of editing, of montage, can be understood, but without assuming that it is the fulfillment of the cinema as an art form nor its radical transformation, a moment either of apotheosis or of demise. Rather, it is an event characteristic of the arts as the technological bases employed to create what we could call, after Margolis, aesthetic utterances. Indeed, the cinema is the site where, and as Margolis suggests, perhaps the most dynamic possibilities are available at this moment in the history of art, such possibilities, though, understood as the general aesthetic possibilities provided by any art. All art is subject, though, to the effects of technological change, which alter, expand, modify the kinds of utterances that can be created, part of a more generally human power to change and be changed, Margolis argues. "The arts—notably film, in our time—provide the principal milieus," as Margolis says, "for deliberately experimenting with such change, incorporating as well as can be done the conceptual and imaginative resources of every cultural sensibility that can be captured and transformed by our own sensibility."[104]

***

To understand the cinema as an art, then, and to understand specific films as works of art, is necessarily, I would argue, to assume that there is some broader category—"art," or "work of art"—to which the cinema, to which films, belong, whether or not that concept can itself in turn be explained or described in some perspicuous manner. As a part of this concept, though, at the very least, there must be some way of explaining change in the history of art, such change often taking the form of a change in technology with

102 Ibid., 126.
103 Ibid., 126–27.
104 Ibid., 127.

the resultant possibility of changes in technique. Indeed, one must be able to explain how, in conditions of almost constant change, technological change specifically, but also cultural change more broadly, works of art are created, which differ often radically in form and style at different times, but which still are what we want to call "works of art." The same must hold for works of film art. I insist upon this as a matter of principle. If we were to countenance the possibility that any particular change to the physical basis of any particular art form could potentially entail a change to the "ontology" *of* that art form, understood as some essential defining and inherent quality, we are left in a position of considerable conceptual difficulty, indeed confusion, such that we would be unable, in any meaningful way, to even talk about art, much less identify cinematic works as works of art, as particular artistic instances, that is, of a particular art form. We would have to confront the possibility of an endless variety of not just different *kinds* of art—that is, according to familiar distinctions, works of literary art, works of sculptural art, works of cinematic art, and so on—but of *ontologically* distinct art forms, that could appear or disappear at any moment, as a result of some particular technological change. This is not only potentially confusing, but it seems to me that it leads us into the realm of essentially arid debate, given that we would always have to be wondering if any particular technological change to any particular art form is "ontological," in what I now want to say is an erroneous sense of the word—if, as a result of some particular change, we are now confronted with some new kind of art, some wholly new form of art. I cannot see how we would ever be able to settle upon the proper criteria for making such a judgement. Indeed, the common claim that the ontology of the cinema has changed given that it is no longer photographic but digital strikes me as a wholly arbitrary declaration. Why is this change, and not any other change (there have been so many in the history of the cinema), the one that changes the cinema *itself*—its *ontology*? What about this change makes such a difference that we want to say that there is a new ontology of film art, that the identity of the cinema has been redefined, that the cinema has relinquished some specific aesthetic capacities replaced by other, different capacities?

Well, the most common answer, as I have already suggested, is *intention*. New digital technologies are supposed to have granted filmmakers a wholly new degree of intentional control. This is what is supposed to distinguish the specific changes wrought by digital technology from any of the many other technological changes in the history of the cinema. This is, indeed, the source of what Rodowick describes as the new ontology of digital simulation and digital editing, what makes new modes of digital compositing "ontologically strange or curious. What appears to be photographic, and therefore causal, is simulated and therefore intentional."[105] Given that all aspects of cinematic representation are now amenable to alteration and manipulation, to direct control as discrete elements, available as such through the interface of the computer screen with digital editing systems, a transformation on the order of ontology is supposed to have occurred, indeed fulfilling the "seamlessness" that has been often thought to characterize editing in the first place. That fact has now been made a comprehensive feature of cinematic composition, the argument goes. "Composited outputs," argues Rodowick, "appear 'seamless' or perceptually real because all the elements now belong to the same ontological order—that of symbolization and its openness to intentional acts."[106] Yet, I will argue, there is no such distinction to make, if we accept, on principle, that the cinema, if it has an "ontology," has the ontology of art in general, that if films, as specific instance of cinematic art, are to be understood ontologically, they share the same ontology as any other work of art.

In fact, I share with Genette some basic misgivings about the very word "ontology," or at least the typical use of the word in the realm of aesthetics. It is, he says, "a cumbersome term," arguing that "one does not seem to me well advised to treat the artistic or nonartistic character of an object as an 'ontological' feature, thus placing the whole of the theory of art under the banners of ontology."[107] He admits that, in order to provide an informative account of art, and of the work of art as a phenomenon, one must begin with what is usually arranged under the "banners of ontology." The first question, he says, approaching the problem of the concept of the "work of art," and, as he admits, "to put it in Scholastic terms, is that of the so-called ontological status of works of art—more simply, that of their mode, or, doubtless, better, their 'modes of existence.'"[108] Genette accepts that the word "ontology" (in noun form) has become common in many accounts of art and art works, which are so often considered (adjectivally) in their "ontological" constitution, commenting on "the (surprising) use of this adjective in contemporary analytic aesthetics."[109] It is "surprising" to him, in this context, given its value in much so-called "Continental" philosophy,

105 Rodowick, *Virtual Life of Film*, 170.
106 Ibid.
107 Gérard Genette, *The Work of Art: Immanence and Transcendence*, trans. G. M. Goshgarian (Ithaca: Cornell University Press, 1997), 6, n. 10; 7.
108 Ibid., 6. Genette takes the phrase "modes of existence" from Étienne Souriau, *La correspondance des arts: Éléments d'esthétique comparée* (Paris: Flammarion, 1947).
109 Genette, *Work of Art*, 6, n. 10.

with its often "metaphysical" connotations. It is, I will observe here, as common in film theory, in both its "continental" and "analytic" forms. Genette admits that the question likely cannot be avoided, though, at least not until a host of other conceptual problems in our accounts of art are addressed, but he advocates for a "restricted ontology," aiming, as he says, "to define 'mode of existence' as narrowly as possible; or—but this comes to the same thing—I want to situate as far outside the realm of the ontological ('what is X') as possible everything I consider functional ('what is X used for?' or 'How does X work?')."[110]

It is in this respect, under the restricted terms of the *functional*, as a part of, or in contrast to, what has traditionally been understood as the "ontological," conceived as the quality produced as a result of some essential nature, or substance, that I have undertaken to consider the aesthetics of digital montage, the mode of existence of works of film art in the digital era, as the products partly of an aesthetics of digital montage. As a necessary corollary of his restricted ontology, emphasizing function over essence—"how does" over "what is"—the "work of art," for Genette, is characterized by a thoroughgoing intentionalism. I quote again his definition of a work of art, which is, though, as he insists, a "provisional definition," which he says is to be regarded as "hypothetical," pending further elaboration: "*a work of art is an intentional* [intentionnel] *aesthetic object*, or, what comes to the same thing, *a work of art is an artifact* (*or human product*) *with an aesthetic function.*"[111] These are effectively equivalent formulations, he says, "because an intentional object can be only a human artifact, unless one puts forward a theological hypothesis (which I refrain from doing) about the divine intentions embodied in natural things."[112] Genette explains further:

> The two formulations are equivalent, again, because *aesthetic object* means "an object in a position to produce an aesthetic effect," while "function" means "intentional effect." The inclusion of both *artifact* and *function* in our definition might be deemed redundant, inasmuch as there are no artifacts without functions (human beings do not produce anything without functional intent), but the function of an artifact is not always aesthetic and is not often exclusively aesthetic; indeed, an artifact's functions do not always include an aesthetic one. On the other hand, an artifact without an (intentionally) aesthetic function (an anvil,

110 Ibid., 7

111 Ibid., 4; emphasis in original.

112 Ibid., 4–5.

> for example) can produce an unintended aesthetic effect, which is to say that it can be perceived [*reçu*] as a (simple) aesthetic object, much as a natural object—a flower for instance—can.[113]

I quote Genette at some length here, as I find this perhaps the most perspicuous and effective account of what we would want to call a work of art, precisely insofar as it is reduced to the basic fact of intention, but without excluding the possibility of aesthetic pleasure which may indeed be found in such objects as anvils (made without *aesthetic* intention) and flowers, as he says, or similar phenomena, like snowflakes (made without intention at all). The confusion on these basic matters is the source, I will suggest, of some considerable confusion in film theory, which very often, and as Genette puts it, considering the arts more generally, tends often to treat either the "artistic or nonartistic character of an object as an 'ontological' feature." This is manifested, I will suggest, when, for instance, and to use one of the very examples offered here, André Bazin says that the "ontology of the photographic image," is the result of the fact that photography, "affects us like a phenomenon in nature, like a flower or a snowflake, whose vegetable or earthly origins are an inseparable part of their beauty."[114] This is indeed to confuse artistic and nonartistic features, and to confuse the intentional ontological quality (which is the "aesthetic function," equivalent to the "intentional function") of an artwork, made contingently out of photography, with the ontological quality of a flower, which may indeed be appreciated aesthetically, but which does not inherently possess such a quality as a *function*.[115] Genette is in fact discussing the so-called "readymade," the complex and puzzling kind of object created so (in)famously by Marcel Duchamp, when, for instance, he simply presents, in more or less "readymade" form, an object like a urinal, purchased at a plumbing supply store, or a snow shovel or a bottlerack, purchased at a hardware store, as an aesthetic object. As described, most significantly, by Arthur Danto, as Genette says, such works are presented as manifesting an "ontological" duality, which is not in the end logically reconcilable. Danto, Genette says, "describes the difference between two physically indistinguishable objects, of which one

113 Ibid., 5.

114 André Bazin, "The Ontology of the Photographic Image," *What Is Cinema?*, 13.

115 At least not as an inherent function, as Bazin implies. A flower, may of course, be submitted for aesthetic consideration, when placed in a pot, for instance, on one's doorstep, but what matters then, it must be stressed, are not its "vegetable or earthly origins," but its intentional placement, in said pot or on that doorstep.

is an artwork (for example, Duchamp's *Bottlerack*), while the other is not (an identical bottlerack in the basement of BHV)."[116]

Genette's point is that such a distinction is effectively unnecessary. There are not two objects in question, if we are concerned with the work by Duchamp as a work of art, because Duchamp has simply taken an object (*one* object), made with a specific intention, with a particular function in mind, and merely offered it up to perform another function, offered it, that is, with an aesthetic intention (as a "candidate" for aesthetic appreciation, which the recipients of such offer may or may not, of course, accept). There is no ontological mystery, as Danto seems to imply. Similarly, I would argue, there is no ontological mystery, as Bazin implies, when photographs, which he suggests lack precisely an *intentional* quality, which are supposed to affect us like a flower, or a snowflake, are used in the creation of a work of film art. The implication of his argument is that there are two kinds of things, the photographic image (the natural thing, created causally, affecting us like a flower) and the work of film art (which is intentional, except for that part that is affecting us like a flower), when in fact, if we are to accept Genette's reasoning, it is merely a matter of function. There is, in both cases, only one thing. A photograph may of course be experienced in terms of its inherently *causal* qualities, as an object that comes into being through specific non-intentional processes, but that is to be distinguished from the ways that a photograph may then be used intentionally, as the material, for instance, of a work of art, a film, which must then be judged entirely on the basis of a presumed intentionality, with which the entire work is imbued, according to the now intentionally *aesthetic* function it is being made to perform (under which any of the non-intentional or "natural" qualities of the material from which it is made must be subsumed).

I cannot offer any more argument than this, but insofar as I pursue a *functionalist* account of the cinema, as an art form, thus construed, and of editing, of digital montage, as a set of techniques (that may be) used in the creation of films, of works of film art, it is based in a restricted ontology of *intention*, as proposed by Genette.[117] It is in this respect, as I am arguing,

116 Genette, *Work of Art*, 7. The translator provides an explanatory footnote, noting that BHV is the common abbreviation of the Bazar de l'Hôtel de Ville, the department store in Paris, with a hardware section, where Duchamp in fact bought the bottlerack which he then presented as a work of art. See Arthur Danto, *The Transfiguration of the Commonplace* (Cambridge, MA: Harvard University Press, 1981), 3, 41–42, 138–39. This is the reference provided by Genette.

117 I have offered some initial observations on the concept of ontology, as used in considerations of digital editing technology, emphasizing intention over the more common accounts of the ontological status of the cinema as originating in some supposedly non-intentional material,

that editing cannot be distinguished in any principled way from any other technique in the cinema, as either the technique that provides the cinema with its aesthetic identity, as for instance the Soviet montage theorists are supposed to have believed, and as, for instance, Valerie Orpen accepts. It is on this basis that Lev Manovich argues that, with the comprehensive subordination of all aspects of cinematic composition and representation to the logic of editing, to a thoroughgoing "spatial montage," the cinema has finally achieved a fully intentional status, which he celebrates as the achievement of the avant-garde ambitions of the original montage theorists. D. N. Rodowick accepts Manovich's argument, yet with a sense of melancholy, mourning the *loss* of some unique, non-intentional, supposedly "causal" quality of the cinema, now that it has been comprehensively transformed into a thoroughly "intentional" form. Yet, in both cases, an erroneous distinction is established between two different and irreconcilable functions. The cinema, insofar as it performs an intentional function, insofar as it performs, and what amounts to the same thing, an *aesthetic* function, does so according to a general ontology shared by all art, and by all art works.

Much of this will be implicit in what I argue in subsequent chapters, but it is the basis upon which I endeavor to account for the transition from film editing to digital montage. A more comprehensive argument for an intentionalist aesthetics of filmmaking, of the cinema, would be the subject of another book. In everything that follows, though, such an aesthetics will be presumed. The reader will have to judge if that presumption is sufficient to accept the account that I provide. The main way in which I will make such a presumption manifest is by refusing any distinction between editing as a technique and any other possible technique in the cinema, by refusing thereby any distinction between what is often presented as the "raw material" of filmmaking, which is supposed to have been the "photographic image," and the specific techniques with which that raw material is produced, supposedly in some essentially causal or non-intentional way, and the techniques, like editing, which is where any actual intention in the cinema is supposed to be found. According to the restricted ontology that I am applying here, the cinema, if it is to be understood as an art form, must be understood to be imbued, in its entirety, with an intentionality, that is, to be performing an aesthetic function, which is to say an intentional function.

in a short essay. See Marc Furstenau, "Film Editing, Digital Montage, and the 'Ontology' of the Cinema," *Cinémas* 28, nos. 2–3 (Spring 2018): 29–49. I do not mention Genette there, but have since found his account of ontology, or "modes of existence," to be the most convincing explanation and analysis of the necessarily intentional quality of works of art.

It is in this respect, further, that no clear distinction can be made between editing and any other technique in the cinema, insofar as any individual technique is supposed to have its source in any particular kind of technology, in the way that the photograph is supposed to be produced technologically in some non-intentional manner.

4. Digital Montage and the Ontology of the Cinema

Abstract
In this chapter, I consider the theory of *montage*, as developed specifically by Sergei Eisenstein, and it effects on subsequent accounts of the concept of editing as a technique. The most common arguments against Eisensteinian theory are derived from the critical and theoretical works of André Bazin. I discuss the effects of this debate on recent conceptions of editing in the digital era, as they are bound up in debates about the ontology of art and of the cinema.

Keywords: Film theory, technique, technology, aestethics, Eisenstein, Bazin

Arguably, no work on the subject of film editing, and thus no work on digital editing, can avoid contending with the most powerful and influential theory of editing, proposed by Sergei Eisenstein, often described as a "formalist" film theorist, from whom we most directly inherit the term "montage," which for Eisenstein is essential to the cinema as an art. Nor, as I have been suggesting, can one ignore the "realist" theories, according to which editing, or montage, had necessarily to be moderated, in favor of the power of the image, as argued perhaps most famously and influentially by André Bazin, for whom it is, by contrast, photography that is definitive. Eisenstein's theory is difficult, perhaps impossible, to summarize. Certainly, it cannot be reduced to a simple thesis or principle. As Jacques Aumont has said, in his critical analysis of Eisenstein's writings on the subject, "strictly speaking, there is no single theory of cinema (or theory of montage, since for Eisenstein the two are one and the same) to be constructed out of his work."[1] Part of the

1 Jacques Aumont, *Montage Eisenstein*, trans. Lee Hildreth, Constance Penley, and Andrew Ross (Bloomington and Indianapolis: Indiana University Press, 1987), 156.

Furstenau, M., *The Aesthetics of Digital Montage: Film Editing and Technological Change.* Taylor & Francis Group, 2024
DOI 10.5117/9789463722803_CH04

complexity of Eisenstein's theory of editing, as Aumont notes, is that it is indeed construed as a theory of the cinema itself, such that the cinema is reduced to the single and supposedly fundamental technique of editing. Moreover, that technique is imbued in the theory with an expansive power, from which the cinema is thought directly to derive its specific identity. On the basis of Eisenstein's claims, and of "montage" theory more generally, as it emerged mainly from the specific filmmaking strategies of several notable filmmakers in the Soviet Union in the years immediately after the Bolshevik Revolution, Eisenstein most notable among them, editing came to be understood as the substantive basis of the cinema as an art, as the art of montage. It is to Eisenstein, or to montage theory more generally, that one may turn to account for the recent changes to editing technology, as I indeed suggest in my title—although my account of "montage" differs in most important respects from the theory as generally understood.

Against any theory of "montage," there are those who instead insist that it is the image, and specifically the photographic image, that is the source of the cinema's aesthetic identity. It is André Bazin who is most associated with such a view, whose arguments about the "ontology" of the photographic image provide the basis of most accounts of the current state of the cinema, and of new digital editing technologies. Bazin's "theory," too, is difficult to summarize in any straightforward way. Daniel Morgan has argued that its very complexity as a theory has not been fully grasped, that Bazin is that figure in the history of film theory "whose intelligence and insight in grappling with the difficult problems of style and ontology has been misunderstood and therefore mostly rejected."[2] In the digital era, he says, when the "advance in digital technologies of image production and manipulation has challenged traditional views about photography and film," the so-called "classical" film theories, but especially Bazin's, may seem "inadequate and irrelevant."[3] Yet, he argues, if we are to "develop new theories to keep up with a rapidly changing media landscape, it is also important not to forget the ambitions of classical theories."[4] Among the classical theorists, he presents Bazin as perhaps the most significant and still relevant, specifically in terms of his

2 Daniel Morgan, "Rethinking Bazin: Ontology and Realist Aesthetics," *Critical Inquiry* 32 (Spring 2006): 444. For another more recent effort to present Bazin's critical as a coherent theory of film, see Angela Dalle Vacche, *André Bazin's Film Theory: Art, Science, Religion* (Oxford: Oxford University Press, 2020). I offer a critical account of Dalle Vacche's argument, in a review of the book. See Marc Furstenau, "Review of *André Bazin's Film Theory*," *Canadian Journal of Film Studies* 30, no. 2 (Fall 2021): 116–20.

3 Ibid., 443.

4 Ibid.

account of editing, or montage, which, as Morgan notes, Bazin is willing to accept as a technique—even in the form of what Morgan calls "internal montage," whereby compositing techniques are used, as in many scenes from one of Bazin's favored films, Orson Welles' *Citizen Kane* (1940), so that photographic imagery is superimposed on matte paintings, for example, to create a synthetic or artificial sense of space. Arguing against too simplistic accounts of Bazin's theory as a theory of "direct realism," Morgan notes that "Bazin does not claim that Welles is being faithful to an antecedent reality," quoting Bazin, who says that in such a case, with such optical effects, "an impression remains of continuous and heterogenous reality."[5] To the extent that montage is used, even "internal" montage—or what Lev Manovich now calls, in the digital era, "spatial" montage—it is acceptable for Bazin, Morgan argues, as long as such an "impression" is maintained, which coincides with what Morgan presents as Bazin's account of the ontology of the cinema. Morgan argues that "while Bazin's realism is oriented by the ontology of the photographic image, it is not determined by that ontology and remains open to a range of styles and genres."[6] Nevertheless, it is that very "ontology," or the specific form of ontological analysis, on display arguably in both Bazinian theory and its putative opposite, Eisensteinian theory, that I seek to counter. Whether "determined" by it, or merely "oriented" by it, the ontology of the photographic image is still at the heart of Bazin's theory, and in recent defenses of it, and is unwarranted precisely insofar as it is conceived of as opposing the imaginative and intentional efforts of the filmmaker to the strictly causal force of the photographic process, understood to be itself the source of some autonomous aesthetic value. Similarly, in Eisenstein, and Eisensteinian theories of editing, montage is understood as that technique that in fact overcomes or properly subordinates the photographic ontology of the cinema, understood as mutely causal, in favor of an aesthetics of manipulation. In such terms, though, it effectively reiterates the claim that the causal power of the photograph is what is in fact arrayed against artistic intention, and which, in this case, must simply be overpowered, forced into performing an aesthetic function, conscripted, so to speak, into artistic service.

In both cases, editing is indeed understood as a necessary and perhaps inevitable aspect of cinematic representation, but, from the Bazinian perspective, it is supposed to be subordinated to the fact of recording, which places appropriate constraints upon the techniques of montage. From this

5 Ibid., 456. Morgan is quoting from André Bazin, *Orson Welles: A Critical View*, trans. Jonathan Rosenbaum (Los Angeles: University of California Press, 1991), 77.

6 Morgan, "Rethinking Bazin," 458–59.

perspective, the recent changes to editing technology have compromised the integrity of the image, freeing editing from all constraint. From the Eisensteinian perspective, digital editing is what has finally allowed for the artistic liberation from the constraining force of the photograph. In both cases, from either perspective, the view is that the recent changes to the technical apparatus for film editing have transformed the identity of the cinema, even if that identity, if the so-called ontological status of the cinema, had been understood in two apparently irreconcilable ways. I will consider the two positions, comparing them, in order to establish the basis for another kind of ontological analysis, potentially reconciling and overcoming the long-standing distinction between image and montage.

***

This central claim from Eisenstein, that the cinema is reducible to the fact of editing, or montage, has been widely accepted, if often only implicitly, in many subsequent accounts, such that the latest changes to the technical apparatus of editing are thought to have changed the cinema itself, to have altered the very identity of the cinema, to have transformed its ontology. The question, of course, is in what respect? For Eisenstein, the basic artistic and aesthetic potential of the cinema was supposed to lie in its unique capacities for editing. From this perspective, the advent of digital editing systems could be understood as the fulfilment of the cinema's aesthetic progress, the realization of its artistic identity, as an art that has now harnessed the total power of montage, extended to all aspects of cinematic representation, every element of the composition governed by a generalized aesthetics of digital montage, understood as the power of comprehensive control and manipulation. This is the process as described by Lev Manovich, who, developing the accounts of Soviet film theory, considers the advent specifically of digital editing systems, and of a digital cinema generally, as the realization of the avant-garde ambitions of Eisenstein, and of other figure such as Dziga Vertov, whose concept of "kino-eye" is described as an incipient effort to establish full compositional control over cinematic representation. In this view, the creative freedom of editing was contrasted with the relative fixity of the photographic imagery, a quality that has more recently come to be described as its "indexicality," understood as a direct, physical link to reality, as a mode of representation comparable to a physical trace, like smoke from fire or footprints in sand. The early montage theorists, Manovich says, "revolted against the indexical identity of cinema," and were "working against 'normal' filmmaking procedures and the intended uses of film technology."[7] He argues that the "general effect of the digital revolution is that avant-garde aesthetic strategies came to be embedded in the commands and interface metaphors of computer software. In short, *the avant-garde became materialized in a computer*."[8] The consequences are significant, he says, arguing that "what used to be exceptions for traditional cinema have become the normal, intended techniques of digital filmmaking, embedded in technology design itself."[9] For Manovich, the digital cinema, and specifically the new digital editing software, have effected a radical transformation, and he argues that "the differences between cinematic and synthetic realism begin on the level of ontology."[10] The cinema, he insists, has become fully susceptible to deliberate manipulation, on par with the other arts, and in line with the most significant avant-garde ambitions. He says that:

> The mutability of digital data impairs the value of cinema recordings as documents of reality. In retrospect, we can see the twentieth-century cinema's regime of visual realism, the result of automatically recording visual reality, was only an exception, an isolated accident in the history of visual representation, which has always involved, and now again involves, the manual construction of images. Cinema becomes a particular branch of painting—painting in time. No longer a kino-eye but a kino-brush.[11]

If, for Manovich, from his montagist perspective, the advent of a digital cinema, and specifically the new powers afforded by digital editing systems, effectively fulfill the cinematic ontology, for others it is a distinct threat against it. The new technology is typically characterized as increasing and expanding the power of editing. Dudley Andrew, from an opposite imagist perspective, has very concisely described the new condition, such that editing is now the processing of "digital information that is then manipulated up to the point of the final cut."[12] For Andrew, this is an aesthetic challenge, a potential threat to the ontology of the cinema, the importation of an alien technology, at odds with the essential qualities of cinematic representation. Andrew accepts the axiomatic proposition of Serge Daney, the French film

7 Lev Manovich, *The Language of New Media* (Cambridge, MA: MIT Press, 2001), 306.
8 Ibid., 306–7; emphasis in original.
9 Ibid., 307.
10 Ibid., 196.
11 Ibid., 307–8.
12 Dudley Andrew, *What Cinema Is! Bazin's* Quest *and its Charge* (Malden, MA: Wiley-Blackwell, 2010), 51.

critic and former co-editor of the influential *Cahiers du cinéma*, quoting Daney, who states, "The *Cahiers* axiom is this: that the cinema has a fundamental rapport with reality and the real is not what is represented—and that's final."[13] Andrew declares that he will accept this axiom, and "throw it against an overconfident discourse of the digital."[14] On the basis of this principle, he argues that the cinema begins with the process of recording, the basic product of which, he insists, should set the essential parameters of any subsequent editing, any subsequent manipulation, certainly against an overweening power of digital manipulation. "Against the all-powerful computer," he says, "traditionalists hold up the camera as a unique device that captures the visual configuration of a given moment, perhaps revealing its truth."[15] This is, he says, basically the view of Bazin, the original founding editor of the *Cahiers du cinéma*. Arguing against Eisenstein, Bazin famously described what he called "the virtues and limitations of editing."[16] The basic principle for Bazin, as Andrew says, is that "the imaginary world shown on the screen must have the spatial density of reality."[17] From this follows an essential corollary. "Editing," says Bazin, "can be used ... only within precise limits, on pain of threatening the very ontology of the cinematic fable."[18]

It is striking that the debates of classical film theory, specifically the debate between the formalists, as represented by the Soviet montage theorists, most prominent among whom was Eisenstein, and the realists, the most significant of whom is undoubtedly Bazin, are what have provided the basic parameters for accounts of the digital cinema, and of digital editing specifically. Andrew is perhaps the most prominent advocate of a Bazinian realist theory of cinema, and his account of editing in the digital era is presented as part of a multi-pronged defense of what he describes as a "cinema of discovery," the idea of which he endeavors to preserve, in the face of its increasingly dominant counterpart, which he rejects, what he calls a "cinema of manipulation." Andrew's summary of the basic Bazinian principle is that there is a tension between imagination and intention,

13 Quoted in ibid., 5. The original quotation is: "L'Axiome *Cahiers*: c'est que le cinéma a rapport au réel et que la réel n'est pas le représenté—et basta." See Serge Daney, *L'exercice a été profitable, monsieur* (Paris: POL, 1993), 301.

14 Andrew, *What Cinema Is!*, 5.

15 Ibid.

16 André Bazin, "Editing Prohibited," in *What Is Cinema?*, trans. Timothy Barnard (Montreal: Caboose, 2009), 75. The more familiar English translation is with the title, taken from the phrase in the essay, "The Virtues and Limitations of Montage," in *What Is Cinema?*, trans. Hugh Gray (Berkeley: University of California Press, 1967), 41–52.

17 Bazin, "Editing Prohibited," 80

18 Ibid., 80–81.

on the one hand, and the "recalcitrant chunks of recorded reality," the photographic image, on the other.[19] This is at odds with montage theory, he says, specifically as articulated by Eisenstein, for whom intentional manipulation, primarily through editing, but through other means as well, was emphasized precisely in order to overcome photographic recalcitrance. Yet, as Andrew notes, Bazin was in fact willing to acknowledge an important degree of respect for the photographic image in Eisenstein's theory and practice, given the more essential realism of his basic imagery, given the real settings and real people he used in his films. Andrew quotes Bazin, who had said that this aspect of Eisenstein's theory, and especially of his filmmaking, his respect at least for the "tension" between intention and photography, is what had been lost in later Soviet theorizing and cinema. Bazin observed that, "because the Soviet cinema was too forgetful of this [tension], it slipped in twenty years from first place to last among the great film-producing nations."[20] For Andrew, it is just such "forgetfulness" that characterizes the digital era, which emphasizes intentional design over the causal reproduction of reality to too great a degree. In contrast, Manovich argues in favor of an aesthetics of manipulation, which he says the montage theorists, and especially Eisenstein, did not pursue doggedly enough, but which is indeed realized now with digital technologies. Manovich describes the potential in Eisenstein's theory of montage, noting that "Eisenstein already used the metaphor of many-dimensional space in his writings on montage," but submits that "his theories of montage ultimately focused on one dimension—time."[21] For Manovich, Eisenstein had been unable to see the full potential of montage, of an aesthetics of manipulation, which has only been revealed now with digital technologies. "The new logic of the digital moving image," says Manovich, "contained in the operation of compositing runs against Eisenstein's aesthetics with its focus on time."[22] Digital montage, what he calls digital compositing, Manovich argues, "makes the dimensions of space ... and frame ... as important as time," such that, "if film technology, film practice, and film theory privilege the temporal dimension of a moving image, computer technology privileges spatial dimensions."[23]

19 Andrew, *What Cinema Is!*, 30.

20 Quoted in ibid. See André Bazin, "An Aesthetic of Reality: Cinematic Realism and the Italian School of the Liberation," in *What Is Cinema? Vol. II*, trans. Hugh Gray (Berkeley: University of California Press, 1971), 25.

21 Manovich, *Language of New Media*, 156.

22 Ibid., 157.

23 Ibid.

A return to the basic questions that had been posed by classical film theory, in both its formalist and realist manifestations, may, indeed, be helpful. Yet, in most instances, some of the basic conceptual problems are merely reproduced, even when the intention, as we see with Manovich, in the case of Eisenstein, and with Morgan and Andrew, in the case of Bazin, is to reveal the enduring relevance of such early theory in the digital era, to review but also revise the theories in light of the most recent developments. In both cases, though, there is still a tendency to reduce the cinema, in its essence, to a particular material or technical fact, and to find the source of the cinema's identity as an art in that fact, which is then made the primary "ontological" fact, against which all other aspects of the cinema must be made to conform. Moreover, such theories are "ontological," in the erroneous sense I have just described in the previous chapter, precisely insofar as they equate forms of cinematic representation with some more basic or elemental forms, namely, for Bazinian theory, "reality" itself, and, for Eisensteinian theory, the "mind," as concepts relating to the basic aesthetic categories of "causality" and "intention." Such conceptual qualities are reduced to, equated with, or understood to derive from, some specific material or formal quality of the cinematic apparatus, from either photographic recording, which captures "reality," or editing, which is supposed to be the manifestation of "mind."

Daniel Morgan argues, against the usual criticisms of Bazin's theory, often dismissed as a naive account of film as the simple and faithful reproduction of the real, that, in fact, "Bazin sees a more complicated relation between style and reality."[24] Yet it is still the concept of "reality" that Morgan argues is at the core of Bazin's theory, as the concept organizing his account of cinematic representation. While Morgan insists that the relation between film and "reality" is not a simply direct one, not to be evaluated in terms of "fidelity," but that it is better described by what Morgan, following Stanley Cavell, describes as "acknowledgement," it is still the case that an understanding of film art is subordinated to an understanding of "reality" itself, as what is somehow more or less properly revealed, represented, or "acknowledged," by a particular style or technique. In this respect, it seems to me, the problem is that the cinema will never be explained in any perspicuous way unless and until we can explain "reality." It is not clear to me why the more local account of a particular art form depends on a solution to this much more capacious philosophical problem. Yet much film theory proceeds on the basis of such an assumption. The "ontological identity" of the cinema, Morgan argues, derives from the fact of photography, which, as a medium, he says,

24 Morgan, "Rethinking Bazin," 445.

summarizing Bazin's view on the matter, "gives us the freedom to form new associations, to have different kinds of relations with the objects in a photograph than we do with the same objects in the world."[25] This is a subtle reinterpretation of Bazin's realism, but still and nevertheless reduces the cinema to a form to be judged adequate only in some relation to the nebulous and vague concept of "reality." As Morgan says, "we can think of Bazin as committed to a specific movement from image to reality. What we learn by watching and responding to films—films that are realist in the appropriate way—can be transferred to the way we engage reality."[26] To say so, though, is to establish effectively normative criteria (Morgan in fact says that Bazin's theory imposes a "moral imperative for films and film criticism"[27]) for judging films as works of art, such criteria derived from the concept of "reality," which is itself never explained in any perspicuous way, and as the kind of "ontological" concept that in fact, and by its nature, one may well argue, defies explanation. To say, then, as Morgan does, that a similar kind of "moral imperative," and a similar account of the complex relations between representation and "reality," is adequate to the new form of a digital cinema, is still effectively to beg the question, to proceed in an analysis of the new form on the basis of an unargued assumption about what reality itself is, to make the formal questions of filmmaking dependent on some broader "ontological" or metaphysical issues.

For Manovich, and despite his efforts to transcend the merely "temporal" emphasis of Eisenstein's montage theory, to include the "spatial dimension," and to thereby fulfill the potential of montage as a technique of intentional manipulation, he nevertheless reproduces some of the basic features of such theory, which depend, again, on certain fundamental and presumably more intentional concepts, most significantly "language," but also "mind." In fact, Manovich is aware of the problem of equating media and mind, which he says is a mistake, and not a new one.[28] Insofar as the supposed

25 Ibid., 454.

26 Ibid., 481.

27 Ibid.

28 There is a long history of thinking that establishes a basic formal relation between film and "mind," as manifested primarily in the techniques of editing. When philosophers like Hugo Münsterberg and Suzanne Langer began considering the film, it was the temporal and spatial elasticity of the form that they argued was the most important. Münsterberg compared the formal techniques of the cinema, like the close-up and what he called the "cut-back," or editing, to specific mental activities, like attention, memory, and imagination. The "photoplay," he said, consists of "moving pictures which, freed from the physical forms of space, time, and causality, are adjusted to the free play of our mental experiences." See Hugo Münsterberg, *The Photoplay: A Psychological Study* (New York and London: D. Appleton and Co., 1916), 78. Langer consider

quality of "interactivity" is thought to be characteristic of new media, he says, this is "just the latest example of a larger modern trend to externalize mental life, a process in which media technologies—photography, film, VR—have played a key role."[29] Manovich describes the "recurrent claims that new media externalize and objectify reasoning, and that they can be used to augment and control it, [which] are based on the assumption of the isomorphism of mental representations and operations with visual effects such as dissolves, composite images, and edited sequences."[30] Manovich resists such "isomorphism," describing the accounts of media processes as the externalization or material, public manifestation of mental processes as a kind of "myth," yet, and insofar as he is concerned to describe the "language of new media," he still reduces a form like the cinema to a more fundamental concept, namely to language, understood precisely as a mental phenomenon. If, as he says, the desire to see in modern, technological media, the manifestation, the making visible, of mental processes, in the hope that in such form mental processes themselves would become susceptible to augmentation or more direct control, that, indeed, we could escape the solipsism of mind to enter some new realm of shared thought, through the visible and publicly shared forms of new "interactive" media, this needs to be shown up for the myth that it is. We are encouraged, he says, to believe that we are directly engaging with mental processes, the mind itself, when we are in fact merely using systems of representation that, while perhaps more powerful and effective than earlier versions, are just representations. Presented with deliberately designed forms, meant to perform a traditionally representational function, Manovich says, "we are asked to mistake the structure of somebody else's mind for our own."[31] Rather, that is, than

the cinema in a short appendix to her book on art theory, entitled "A Note on the Film." Langer describes what she calls the cinema's "virtual" aspects, the most like the "virtual" powers of mind, the result specifically of the possibilities of editing. See Suzanne K. Langer, *Feeling and Form: A Theory of Art* (New York: Charles Scribner's Sons, 1953), 415. Arnold Hauser, in the conclusion to his monumental multi-volume survey of the history of art, considered what was still thought of as the new form of the film, whose novelty consisted partly in its aesthetic distinctiveness. "The most fundamental difference," he says, "between the film and the other arts is that, in its world-picture, the boundaries of space and time are fluid." Hauser makes a basic comparison between film and mind. "The film," he says, "is an elastic, extremely malleable, unexhausted form which offers no inner resistance to the expression of the new ideas." See Arnold Hauser, *The Social History of Art, Volume 4: Naturalism, Impressionism, the Film Age* (New York: Vintage, 1958), 239, 256

29 Manovich, *Language of New Media*, 57.

30 Ibid., 59.

31 Ibid., 61. This "mistake," he says, is like what Louis Althusser described as the process of "interpellation," which itself, as Manovich suggests, depends on a dubious analogy between media

understanding the new "interactive" forms of new media as designed objects, which we should experience as such, as the *product* of another mind, as the manifest results of an intentional effort at design, we are encouraged to imagine that we are making the choices ourselves in our cognitive engagement with the work, as though they are the manifestation of our own thought rather than the product of someone else's. As Manovich says, this "aspect of interactivity remains one of the most difficult theoretical questions raised by new media," and he admits that he will not have a "complete answer" to them.[32] He raises the issue, though, in order basically to dismiss it, in favor of his view of new media as the generalized imposition of the logic of encoding on all modes of representation, including most significantly the cinema. While refusing the claims of media as the externalization of mind, though, he effectively makes a similar claim about the nature of representation, the full capacities of which, he argues, had been suppressed by earlier forms, and which are radically released, in their full potential, by computer technologies. Manifested in a range of "cultural possibilities," as he describes them, whereby, for instance, "animation comes to challenge live cinema, spatial montage comes to challenge temporal montage,"[33] the powers of "new media" are presented as the power to "open culture," to realize representational and expressive potentials that has been suppressed by old forms—and powers that are somehow manifested in the very material of new media, as somehow inherently "intentional," or making a fully-fledged intentionality finally possible, on the unstated but implicit assumption that it was previously impossible, or at least very difficult, in film, glimpsed only by the avant-garde, as opposed, that is, to the inevitably constraining nature of the traditional cinema, still imbued with the bluntly causal, or "indexical," photographic image. "This opening up of cultural techniques, conventions, forms, and concepts," he says, "is ultimately the most promising cultural effect of computerization—an opportunity to see the world and the human being anew, in ways that were not available to a 'man with a movie camera.'"[34]

Yet this has all the problems of most "ontological" accounts, whereby such potential is imagined to be embedded in, manifested in, some particular material form or technical apparatus, assumed to be a constitutive part of

and mind. See Louis Althusser, "Ideology and Ideological State Apparatuses (Notes towards an Investigation)," in *Lenin and Philosophy and Other Essays*, trans. Ben Brewster, (London: New Left Press, 1971), 127–86.

32 Ibid., 56.

33 Ibid., 333.

34 Ibid.

it, to possess qualities and abilities—to open up convention, create a view of the world, to reveal reality, to show us mind, and so on—that are strictly *human* capacities, I would argue, that derive from human action, and that are necessarily possible through *whatever* means are available. While different media certainly provide different affordances, no particular medium provides affordances greater than, or in some broad sense more powerful, more capable of fulfilling the goals of human expression, than any other. Manovich, despite his principled and convincing arguments against the claim that media can somehow make mind manifest, that certain media have the power to make mental processes, cognition or thought itself, available in some sort of tangible, public form, nevertheless makes a similar argument for the inherently creative powers of representation and expression, which are supposed to be fulfilled, or made somehow more realizable, through a new kind of "language," a new kind of procedure for encoding information, making it available for an unprecedented degree of intentional control.

It is the kind of idea that had informed much early theoretical writing on the cinema, which, as a novel form of art, was imagined to be the means that would indeed radically expand the capacities of representation and expression, if not in fact reveal the nature of reality itself, or manifest the very workings of the mind, as the basis or origin of such powers. The cinema, as the form that seemed most capable of recreating perception itself, and in that respect the conceptual powers that derive from the perceptual apparatus, was typically imagined as the basis for a new kind of culture, a cinematic culture, that would fulfill the creative desires thought to be supressed by traditional media. This was arguably at the center of early theories of cinema that emphasized editing as the source of the new power, as in Eisenstein's theory of montage, which was, for him, a material manifestation of cognition. For Eisenstein, as he says in an early statement of his views, "montage (in the technical, cinematic sense of the word) is fundamental to cinema, deeply grounded in the conventions of cinema and the corresponding characteristics of perception," explaining that it is "made up of the juxtaposition and accumulation, in the audience's psyche, of associations that the film's purpose requires, associations that are aroused by the separate elements of the stated (in practical terms, in 'montage fragments') fact, associations that produce, albeit tangentially, a similar (and often stronger) effect only when taken as a whole."[35] As this suggests, and as Jacques Aumont notes,

35 Sergei Eisenstein, "The Montage of Film Attractions," in *Selected Works, Vol. I, Writings, 1922–34*, ed. and trans. Richard Taylor (London: BFI; Bloomington and Indianapolis: Indiana University Press, 1988), 41.

"the Eisensteinian system forges a link between the organization of the film (montage, in a general sense) and the psychic processes."[36] As we shall see, one of the main difficulties in assessing editing, and of accounting for the latest changes to the technology of editing, is the longstanding tendency, so explicit in Eisenstein, but to be found in the work of many others on the subject, to grant the technique such extensive power and significance. Already a challenge to describe simply as a formal technique, the difficulty is only increased once such associations are made with "psychic processes," once some essential relation is established between montage and perception, between film editing and the mind, comparing a cinematic technique to the basic process of comprehension itself.

For my purposes here, though, it is not important to debate such general matters any further. Eisenstein's value as a theorist in fact lies less in the broad claims that he often made, and more in the narrowing of the basic problems to be explained by theory, in his drawing our attention to the specific nature of the problems. In one of his later essays, Eisenstein helpfully reduced the problem of editing to a single, pointed question: "Why do we edit film at all?"[37] He asked this in the context of a fierce debate in the Soviet Union at the time about the value of editing, as the most prominent advocate of montage theory, and in the face of the ideological charge against him and others of "formalistic" excesses.[38] Following the Bolshevik Revolution, during an initial period of radical artistic freedom, Eisenstein and other filmmakers in the new Soviet cinema primarily emphasized the power of editing, conceived of as "montage." There were experiments being undertaken in almost all the arts. New forms were being created, new methods tried, with the hope that a novel, revolutionary kind of art would emerge. The cinema, as a new art itself, seemed perhaps the most promising in such a context. For the filmmakers at the time, the most significant artistic possibilities for cinema were to be found in editing, in the conceptual and expressive arrangements of shots, through the free juxtaposition of images, to create

36 Aumont, *Montage Eisenstein*, 195.

37 Sergei Eisenstein, "Montage 1938," in *Sergei Eisenstein: Selected Works, Vol. II, Towards a Theory of Montage*, eds. Michael Glenny and Richard Taylor, trans. Michael Glenny (London: BFI, 1991), 296. Written between March and May, 1938, the essay was first published, in Russian, in the journal *Iskusstvo Kino*.

38 Eisenstein's difficulties with the official cultural organisations, which by the 1930s were attacking "Left cinema" in the U.S.S.R., are described concisely by David Bordwell, who notes that, for instance, after their release, "*October* and *Old and New* were charged with 'formalistic' experimentation," which was thought to make them incomprehensible to a mass audience. See David Bordwell, *The Cinema of Eisenstein* (Cambridge, MA: Harvard University Press, 1993), 22.

complex symbolic, metaphorical meaning to rival anything in the other arts—perhaps even to surpass them in sheer aesthetic power. This initial period had abruptly come to an end, though, and in filmmaking it meant an end to the experiments in editing, as montage theory was brought into question, thus prompting Eisenstein to mount a defense of the technique, to explain why we even edit a film at all. An era of dour "socialist" realism had come to replace the exciting early phase of artistic adventure and exploration, which had been a part of a more general revolutionary fervor.[39] The often-bold experiments in film form, especially the experiments in editing by the montage filmmakers, would come to be seen as elitist and obscure—certainly not "realistic" according to the new (crude) standards for a "socialist realism" that were being imposed. One of the main targets was the non-realistic use of editing, as Eisenstein notes. "There was a period in Soviet cinema," he says, "when montage was proclaimed 'everything.' Now we are at the close of a period during which montage has been regarded as 'nothing.'"[40]

Eisenstein had long argued that editing, or "montage," was indeed the essence of cinematic expression, the technique that would make film the most powerful new form of communication, suitable to an age of mass audiences in a new revolutionary society. "We must," Eisenstein had said, in an exemplary instance of this claim, "look for the essence of cinema not in the shots but in the relationships between the shots. ... *The expressive effect of cinema is the result of juxtapositions*. It is *this* that is specific to the cinema."[41] This central idea, that the cinema was essentially the art of editing, that montage was "everything," was being reiterated as a kind of slogan. Fellow Soviet filmmaker Vsevolod Pudovkin had made similar claims, arguing that film art was based ultimately in editing. Lev Kuleshov, who had established the Moscow Film School, wrote that "the basic strength of cinema lies in montage."[42] It was, we might say, a widely shared assumption, the only debate being just *how*, exactly, editing provided this strength, or formed this basis. By the time Eisenstein wrote his later essay, though, his own views had

39 See, for instance, David Elliott, *New Worlds: Russian Art and Society, 1900-1937* (New York: Rizzoli, 1986), and Régine Robin, *Socialist Realism: An Impossible Aesthetic* (Stanford: Stanford University Press, 1992).

40 Eisenstein, "Montage 1938," 296.

41 Sergei Eisenstein, "Béla Forgets the Scissors," in *Selected Works, Vol. I*, 79, 80; emphases in original. Published originally in the journal *Kino*, July 20, 1926, as a response to an article by Béla Balázs.

42 Lev Kuleshov, *Kuleshov on Film: Writings of Lev Kuleshov*, trans. and ed. Ronald Levaco (Berkeley, Los Angeles: University of California Press, 1974), 52.

become less strident, and his hopes for montage less extravagant—certainly, at least partly, as a result of the political pressure being brought to bear, but also, I would suggest, as his thinking on the matter generally had matured. He now presented the basic issue in a more explicitly even-handed way. "Since we consider montage to be neither 'nothing' nor 'everything,' we now think it necessary to recall that montage is as essential a component of film-making as all the other affective elements of cinematography."[43] Not any more or less important than any other component, but still, *essential*. But what, exactly, makes it so? "Why," as Eisenstein asks, "do we edit film at all?" The answer, he insists, will not come from a consideration of strictly technical matters. "Even the fiercest opponents of montage will agree," he says, "that it is not because we do not have infinitely long strips of film at our disposal and, being doomed to a finite length of film, are forced from time to time to glue one piece of film to another."[44]

Very significantly, if a little facetiously, Eisenstein is rejecting any essentially technical or material analysis of the cinema as the basis for an account of its stylistic or expressive possibilities. He is offering what could be described as a more functionalist account of editing specifically, of montage, and of cinematic expression generally. For Eisenstein, montage is much more than a merely technical necessity. It solves more than the strictly material or physical problem of the necessarily finite length of the filmstrip. It has a fundamentally aesthetic and expressive value—as do other cinematic techniques, though, he now seemed willing to admit, all of which contribute to an overall artistic or aesthetic function. Eisenstein emphasizes the specific function of editing, of montage, but now seems to acknowledge that it is indeed comparable to all the other techniques of filmmaking, indeed to artistic techniques more generally. Describing editing, he emphasizes "its *cognitive* role, which every work of art sets for itself: *the function of providing a coherent, consistent exposition of the work's themes, plot, action and events*."[45] Montage, in this respect, is one of the many techniques available to the film artist, a particularly important technique, but which is in service to the artist's broader aesthetic goals, and whose significance is inseparable from the many other techniques of cinematic representation. That montage was also strictly necessitated by a particular technical limitation, specifically by the finite length of the filmstrip, is merely contingent. The implication of Eisenstein's argument is that, even

43 Eisenstein, "Montage 1938," 296.
44 Ibid.
45 Ibid.; emphasis in original.

if filmstrips were *unlimited* in length, editing, or "montage," would still be an important, perhaps even indispensable aspect, maybe even an essential component, of cinematic expression, understood to perform a more broadly aesthetic function comparable to other techniques. Nevertheless, while he had become willing to accept that the other techniques of filmmaking were as important, editing, or montage, for Eisenstein, was still *essential*, concluding that, even in the new era of state-mandated realism, and perhaps even admitting to some earlier "formalistic" excesses, "skilled practitioners of cinematic art must master all the subtleties of the creative use of montage."[46] This would presumably be the case, he implies, even if the filmstrip were to become unlimited in length, even if the basic, generative, technical fact that made editing necessary in the first place were to be removed as a constraint.

Well, Eisenstein's question takes on a new urgency. Why edit now? It is not, any longer, simply because the filmmaker is "doomed to a finite length of film." In the digital era, the specific technical limitation Eisenstein described, the particular constraint imposed by the finite length of the filmstrip, may well have been overcome. It is quite simply no longer the case that, at some point, given an inherent finitude, working with finite strips of film, that one *must*, at some point, glue one piece of film to another. The length of the "filmstrip" has become, if not infinite, certainly much longer, and the specifically technical characteristic of the apparatus of editing may seem to have more direct bearing now on style than Eisenstein would have been willing to admit. Of course, this is because the actual material object that is being edited is, in fact, no longer an actual strip of *film*. Most "films" are now shot on digital cameras, or film recordings are converted to digital information, which is stored on computers, on a hard disk, available to be retrieved, through the logic of random-access, and then immediately displayed and easily manipulated on a computer monitor. We could in fact say that it is not in terms of "length" that a filmmaker even thinks any more, but rather "space," the amount of storage that is available on a computer hard disk for visual and audio recordings, encoded as information, space that is now, at least for most practical purposes, effectively unlimited. It had been the case that only so much film could fit into a camera, into the magazine attached to it, such that editing was, as Eisenstein admitted, inevitable at some point, given this strictly technical constraint that could not be overcome. Now, though, the recording space is no longer as limited, no longer governed by the *length* of film, but radically expanded by computer storage *space*. Editing, as a result, and somewhat paradoxically, has in

46 Ibid., 326.

fact become *dispensable*, perhaps no longer the essential component that Eisenstein had understood it to be. "Why do we edit film at all?" Well, we may not have to any longer. Digital "editing" seems to promise new stylistic possibilities as a consequence of this liberation from a particular technological constraint, which could well mean, in effect, a liberation from the very need to edit. Despite Eisenstein's insistence, that is, that editing is more than a merely technical requirement, that it is an essentially aesthetic aspect of filmmaking, it may now be the case, with the latest changes to the tools of editing, that editing itself is technically obviated.

At the very least, editing may well be reduced to the status of wholly optional, indeed entirely dispensable, stylistic technique. If, as Eisenstein had insisted, editing was not the direct result of the simple technical fact of the lengths of strips of film, but an artistic technique that performed an aesthetic function, regardless of technical necessity, that function has now been brought into question, partly as a result of the fact that it has become much easier to *not* edit. Indeed, since the advent of digital recording, a significant number of single-shot films have been made, even made in a single take, exploiting the specific possibility provided by new digital cameras, which is to shoot continuously for effectively as long as one wants, for the entire length of a typical feature film, if one chooses. Given the challenge of shooting in a single take, though, many of the recent apparently single-shot films are in fact created with new digital editing systems, which have made it possible to easily disguise cuts, to obscure the editing, and to create the impression of a single, fluid, moving shot. These are still relatively rare, but new technology is being used now to create much longer shots in films, which are, though, again, often constructed through digital editing, consisting of several shots blended together that only appear to be a single shot. Are these stylistic choices the direct consequence, though, of the introduction of the new technology? Does style derive directly from a technical innovation? This is a common, if often implicit, assumption, which Eisenstein himself had subscribed to, but is apparently rejecting in this late essay. Recent changes to the technical apparatus of the cinema, and specifically to editing, though, do seem to be having clear stylistic consequences, profound effects on film as an art, on its very aesthetic identity. As *editing* becomes unnecessary, merely a choice rather than a necessity, *montage*, understood as the expressive, rhetorical use of editing, may lose its stylistic relevance. The very nature of editing, of montage, as a supposedly inherent or definitive aspect of film art is in question. Most significantly, perhaps, editing, the new form of *digital* editing, can be used now to effectively disguise or obscure the very fact *of* editing, creating what could be described as a kind of aesthetic conundrum, or artistic paradox.

***

Editing will at least, it seems, and as many have argued, have to be re-defined. This is, however, a rather subtle and complex matter, such complexity in fact obscured by the tendency to restate the debate simply in terms of classical film theory. On the one hand, editing seems to be more easily dispensable now than it has ever been. On the other hand, though, it may be argued that editing, or digital montage, has in fact become the ubiquitous feature of all filmmaking. It has become much easier to shoot very long takes, to create very long shots, even comprising whole films. These are, though, in most cases, the result of a kind of editing, a literal kind of invisible editing. Separate shots are joined seamlessly to create the impression of a single shot, a fulfillment, in a certain respect, of the idea of "invisible editing." The term "invisible" has usually been used to describe a style of cutting, the so-called continuity style, such that the editing, which does exist and can be seen, *seems* invisible, but which is really only easily overlooked. Now, though, new digital editing systems allow the cuts themselves to be entirely effaced, so that they are truly invisible. Moreover, given that the material itself being edited is now most commonly in digital form, a digital image, rather than the traditional photographic recording, *any* element of that material can be manipulated, any part of the image, according to a much more expansive notion of editing. In this respect, editing, or montage, is now thought to be a pervasive fact of filmmaking, thoroughly comprehensive, extended to the arrangement and manipulation of all elements of cinematic representation. Editing is no longer limited merely to the linear sequencing of shots, but to the arrangement and manipulation of the elements of the shots themselves.

That there has been a significant change to the technology of editing in the cinema is undeniable. Editing has been transformed from the simple act of cutting strips of film into smaller pieces and gluing or taping them together to the now more technically complex process of arranging shots and of potentially modifying the content of those shots with software on a computer. What is puzzling, though, is that, while the *technical* changes to film editing in the digital era have been so significant, while the materials and tools used by the film editor have been so comprehensively transformed, little about editing as an *aesthetic* fact seems to have changed at all. This discrepancy, as I discussed earlier, has been noted by, for instance, Maraj Dhir and Sergej Ivasović. Still, as Dhir suggests, there is a kind of "convergence" between the live-action film and animation, such that "mise-en-scène and how it is shaped through editing will achieve ever greater degrees of plasticity."[47] While Ivasović admits that there seemed to be little aesthetic effect on editing *style*, he nevertheless saw new digital editing systems as a considerable technical advancement with possibly more profound consequences. "For the first time," he notes, "the editor may influence and control intrashot structures, i.e., the editor in the process of editing may construct/create shots which do not exist in the recorded material."[48] The changes, as he puts it, may well amount to a "real revolution."[49] As he explains, the "basis of computer digital non-linear editing is the concept of *true random access*, which refers to the ability to access every image of the material at any moment and in any order. Practical implications of this concept to the creative process and final result have been immeasurable."[50] With what he calls "multilayer" editing, "the editor can manipulate the visual structure of a shot … edit within the shot; an entirely new form of editing image expression [is] created."[51] While this had not (or not yet) changed the basic fact of editing as a creative component of the cinema, with the "cut" as its most basic element, Ivasović nevertheless saw considerable creative potential in the new technology, which will, he says, directly affect the cinema, to "enable it to advance in new directions, limited only by our imagination."[52]

Others have made the same point, reaching the same conclusion, and have considered the consequences in more detail. The changes that have been wrought by digital editing systems in particular are thought to be so profound that the very nature, the aesthetic identity, the *ontology* of the cinema is altered—but as "ontology" had been conceived in so much classical theory. Transformed from a fundamentally temporal technique to a spatial one, film editing has become digital montage. This is a part, perhaps the central part, of the overall transformation of the cinema in the digital era, the assumption that the cinema has been changed in a profound or fundamental way. "Computer media," Manovich has said, in one of the more famous and influential versions of the claim, "redefine the

47 Meraj Dhir, "The Modern Entertainment Marketplace, 2000–Present: Editing," in Charlie Keil and Kristen Whissel, eds. *Editing and Special/Visual Effects* (New Brunswick, NJ: Rutgers University Press, 2016), 170.

48 Sergei Ivasović, "Digital Editing," *Moveast*, 7 (2002): 151.

49 Ibid.

50 Ibid., 149; emphasis in original.

51 Ibid., 154.

52 Ibid., 157.

very identity of cinema."[53] Manovich has famously described what he calls "spatial montage," which, he says, "represents an alternative to traditional cinematic temporal montage, replacing its traditional sequential mode with a spatial one."[54] Manovich distinguishes the new digital editing systems from the traditional ones they have replaced, saying that "the differences between cinematic and synthetic realism begin on the level of ontology."[55] A digital cinema is novel just insofar as editing has become spatial rather than temporal. It is the change to editing that is the most consequential change. Manovich describes a "new cinema in which the diachronic dimension is no longer privileged over the synchronic dimension, time is no longer privileged over space, sequence is no longer privileged over simultaneity, montage in time is no longer privileged over montage in space."[56] This is the specific change that is at the heart of the broader changes affecting the cinema in the digital era. "As traditional film technology is universally being replaced by digital technology," Manovich says, "the logic of the filmmaking process is being redefined."[57] If it had been the case that "live-action footage were left intact in traditional filmmaking, now it functions as raw material for further compositing, animating and morphing," the result of which, he argues, is that "film obtains a plasticity that was previously only possible in painting or animation."[58] In the past, he says, "editing and special effects were strictly separate activities. An editor worked on ordering sequences of images; any intervention within an image was handled by special-effects specialists. The computer collapses this distinction."[59] With this collapse, the status of editing at once loses its individual identity and becomes radically expanded in its powers. Manovich argues that "reordering sequences of images in time, compositing them together in space, modifying parts of an individual image, and changing individual pixels become the same operation, conceptually and practically."[60]

It is, though, just such a collapsing of practical and conceptual matters that I question, against which I have been arguing. It is not clear to me that changes to the practical, or *technical*, aspects of filmmaking necessarily have such conceptual, or *aesthetic*, consequences, that even such significant changes

53 Manovich, *The Language of New Media*, 293.
54 Ibid., xxxiv.
55 Ibid., 196.
56 Ibid., 326.
57 Ibid., 300.
58 Ibid., 301.
59 Ibid.
60 Ibid.

to the basic technical apparatus of filmmaking affect the aesthetic identity of the cinema as an art form. Yet this is a common assumption. Moreover, and while Manovich celebrates it as a kind of triumph, this is often seen as a kind of aesthetic loss, a degradation of the cinema as an art form, which is properly limited in its creative and aesthetic potential. Wheeler Winston Dixon has described what he calls "the unreliability of the manufactured image," the altered status of cinematic representation in the digital era, which he presents as an era of rapid technological change and innovation. The most significant effect, for Dixon, is the breakdown of the distinction between editing and special effects. "For all the benefits of these technological advances," he says, "it is impossible to overstate the impact, both in terms of image production and image reception, of digital special effects (and the use of digital production method as a kind of overall 'finishing process' for filmmaking as a general practice) which have effectively (and surreptitiously) changed the language of what we have come to know as 'motion pictures.'"[61] It is as such a comprehensive "finishing process" that the expanded powers of digital technologies are understood, and that "finishing" is what digital editing systems make possible. Indeed, Dixon reduces the digital transformation overall to an account of a change to editing, to what he calls "montage."

Like many film theorists, Dixon contrasts contemporary "digital imaging," presented as a strangely insubstantial process, with the necessarily material nature of filmmaking in the past, which would have required working with physical strips of film, which were to some important degree difficult to alter, susceptible only to the limited manipulations of editing. With the recent technological transformation, though, "digital imagery offers a nearly limitless list of possibilities for the plastic manipulation of the cinematic image."[62] With "digital imaging," he says, "films are now made out of nothing at all, just magnetic impulses entered into computers and then downloaded from discs to create a chroma-key blue-screen world of endless transfiguration and corporeal unreality."[63] This he contrasts explicitly with traditional editing, what he calls "external montage," which was appropriately limited in its manipulative possibilities:

> Image manipulation through external montage is one matter, in which shots are intercut for emotional or associative impact. But Digital Domain,

61 Wheeler Winston Dixon, *The Transparency of Spectacle: Meditations on the Moving Image* (Albany: State University of New York Press, 1998), 22.

62 Ibid., 28.

63 Ibid., 112.

> ILM, and their brethren are creating a new sort of montage—montage within the frame, without the familiar warning, or "limit" signals we have been trained to look for (wipes, matte lines, differences in film stock grain, inaccurate image sizing and other flaws). These visual cues alerted us to the artificiality and constructedness of the manufactured image we see on screen; now these cues are absent. Images within images are nothing new. Images seamlessly bonded to images within the same frame are a different matter altogether.[64]

Dixon does not mention any contemporary digital editing systems, but he identifies two of the most significant visual effects companies, Digital Domain, which was co-founded in 1993 by film director (and one-time effects specialist) James Cameron, and ILM, or Industrial Light and Magic, founded in 1975 by George Lucas.[65] It is such effects companies that are "creating a new sort of montage," emphasizing the degree to which, for Dixon, editing has indeed become part of a more expansive form of digital imaging. No longer a distinct aspect of filmmaking, editing is understood to have been absorbed by and subordinated to the production of special effects, which are increasingly difficult to identify as such given the power of "montage within the frame." Digital editing is thought of at once as a radical expansion of the power of editing, limited no longer to the simple, "external" process of arranging shots and sequences, but also as a kind of diminution, contributing to a loss of editing's distinctiveness and independence as a craft, pressed into the service of an increasingly illegitimate effects-laden cinema of spectacle.[66]

Rather than simply arranging photographed still images into linear arrangements, editing is now understood by Dixon as merely a part of a more comprehensive process of fabrication. Traditional editing is thought of

64 Ibid., 32.

65 For an account of ILM's role in the rise of digital cinema, written by a former employee of the "Droid Works" division of ILM, see Michael Rubin, *Droidmaker: George Lucas and the Digital Revolution* (Gainesville, FL: Triad Publishing, 2006). Rubin is himself an editor, and he has written several guides for using digital editing systems such as Final Cut Pro and iMovie. For a history of Digital Domain, see Piers Bizony, *Digital Domain: The Leading Edge of Visual Effects* (London: Aurum Press, 2001).

66 Martin Lefebvre and I have considered the fraught nature of "special effects" for film critics and theorists, who have either largely ignored effects, which we insist have been an integral part of filmmaking from the very origin of the cinema, or have bemoaned them as illegitimate incursions into the process of filmmaking understood as properly "photographic." See our "Introduction" in Martin Lefebvre and Marc Furstenau, eds., *Special Effects on the Screen: Faking the View from Méliès to Motion Capture* (Amsterdam: Amsterdam University Press, 2022), 7–34.

as the sort of necessarily "flawed" process, "warning" us with its inevitable traces about the "constructed" nature of its products, that has been replaced by processes for the creation of "seamless" illusions, wholly concocted by digital means but appearing to be traditionally photographic. Dixon argues that "one of the most disturbing aspects of the new wave of digital effects films is that they do not seem—at first glance—to contain any effects at all. Philip Rosen has expressed a similar concern, and described what he calls "digital mimicry," the "digital capacity to transform or fake photographs. The compositional flexibility of this imaging technology, which is supposed to be non- or even antiphotographic, includes imitating photographic form."[67] Such power of "mimicry," this "compositional flexibility," is what is thought to be most powerfully manifested in expansive digital editing systems, which extend the power of editing to the shot, to the image itself. The assumption is that the integrity of the shot has been compromised. Indeed, the very concept of the "shot" is brought into question, as perhaps no longer applicable in the era of digital cinema.

As D. N. Rodowick has argued, "there may be 'takes' in digital cinema, but there are no shots. Or rather, what was previously considered a shot has now become a highly variable element open to interactive manipulation at the most discrete levels. In this respect, cinema has become more like language than image, with discrete and definable minimal units (pixels) open to transformations of value and syntactic recombination."[68] Rodowick, as we have seen, has developed the concept of digital montage—of intrashot editing, internal or spatial montage, of editing within the shot—in perhaps the most detailed form, considering the consequences that he sees flowing from the compromised integrity, if not actual disappearance, of the shot itself. In its place—in the place of traditional editing, which had been separate and distinct from the techniques of compositing and special effects—he proposes a new concept, what he calls the "digital event," which, he says, "corresponds less to the duration and movements of the world than to the control and variation of discrete numerical elements internal to the computer's memory and logical processes."[69] He argues that this is a fundamental change to the very ontology of the cinema. While the cinema had traditionally been limited to the photographic recording of "real" events in the world,

67 Philip Rosen, *Change Mummified: Cinema, Historicity, Theory* (Minneapolis and London: University of Minnesota Press, 2001), 309.

68 D. N. Rodowick, *The Virtual Life of Film* (Cambridge, MA: Harvard University Press, 2007), 166.

69 Ibid.

necessarily expressing thereby real duration and real movement, Rodowick is suggesting, the "digital event" is severed from the temporal and historical reality of the world. While cinematic imagery was once created through an "analogous" process of recording, such recordings then arranged through editing, necessarily limited by the material "integrity" of the recordings, such integrity is now compromised, subject to the processing power of computers. "Several aesthetic conclusions may be drawn," Rodowick says, "from this abstract, technological description. First, as constituted through digital capture or synthesis, the image is always 'montage,' in the sense of a singular combination of discrete elements. Even an unaltered digital still is already a work of montage in this respect."[70] Rodowick argues that "the spatial unity of the image in time can no longer be assured or attested to by the digital image," and that "the powers of indexicality are weakened and decentered by the process of digital conversion," such that "the expression of duration is transformed—it becomes other to the powers of film and calls for a new medium."[71] He argues that "the causal chain of analogy is broken," and that "the electronic screen expresses another ontology, which I have characterized as an increased attention to the present and to the control of information."[72]

It is through the means of an expanded form of digital editing, he argues, that such an "ontology" of combination and control is manifested, insofar as the basic integrity of the traditional unit of editing, the shot, is compromised. He considers, as an example of the new "digital event," the 2002 film *Russian Ark*, directed by Alexander Sokurov. The film is famous as one of the first single-shot, single-take films made in the digital era, taking advantage of the specific affordances of new digital camera technology, which allowed Sokurov to shoot the entire film, roughly ninety minutes long, in one take. While traditional film cameras could only hold enough film for about ten minutes of shooting—at the very most enough for about thirty minutes—Sokurov used a camera with an attached computer hard disk with enough storage to capture about 110 minutes of shooting. While consisting of a single shot, the film was subjected to some significant post-production alteration, digital modification, such that, Rodowick argues, the film in fact consists of considerable instances of *montage*—understood now as the arrangement, or re-arrangement, of elements within the shot, available for such arrangement as a result of their being in digital form. Specific

70 Ibid.
71 Ibid.
72 Ibid.

sections of the film, he argues, "illustrate how well the separability of image components into discrete layers brings something fundamentally new to digital images: the possibility of what Manovich calls 'spatial' montage."[73] Rodowick argues that the new composited representations created by digital editing cannot be simply compared to film imagery in its original form, or to editing in any traditional sense. In its place there are processes that allow for the fine-grained control of every aspect of the shot, which, and as opposed to "filmic editing," he calls "compositing" or "digital synthesis." As manifested in a film like *Russian Ark*, Rodowick says that there is now "something like a palimpsestic combination of data layers," that "what is combined in 'montage' is no longer just contiguous spatial wholes as blocks of duration, for any definable parameter of the image can be altered with respect to value and position."[74] The change, he argues, is fundamental. "Digital synthesis produces an image that never occurred in reality; it is a fully imaginative and intentional artifact."[75] The new processes of synthesis and compositing, the new modes of digital montage, he says, combine discrete elements "in a way that produces something ontologically strange or curious. What appears to be photographic, and therefore causal, is simulated and therefore intentional."[76] He argues that it is not merely a matter of style, insisting that "synthetic imagery is neither an inferior representation of physical reality nor a failed representation of the photographic, but rather a fully coherent expression of a different reality, in fact, a new ontology."[77]

***

It is in such accounts that we can see the degree to which the advent of digital editing systems has re-opened some basic debates in film theory, and a consideration of this specific technological innovation is indeed an opportunity to reassess and re-evaluate some of the most essential theoretical assumptions about the cinema. The change, though, as I have been arguing is not as significant as often claimed, and the sense of profound transformation, as expressed by Rodowick, is in fact a consequence of the ways that the cinema has traditionally been understood. An aesthetics of digital montage should instead be understood as part of a rearticulation of

73 Ibid., 169.
74 Ibid., 169.
75 Ibid.
76 Ibid., 169–70.
77 Ibid.

certain basic principles of film theory, of an aesthetics of the cinema more generally, according to which technical change would be understood as an integral and inevitable aspect of the cinema as an aesthetic phenomenon, rather than as a force that changes its identity, that alters or transforms its ontology. To begin, though, it is important to consider the undeniable changes that have indeed occurred to which Rodowick in fact draws our attention. The hallmark of the era of "digital montage," as I have suggested, does seem to be the *dispensability* of montage as a visible stylistic component of the cinema, the overcoming, for better or worse, of a once unavoidable technical necessity, from which aesthetic and expressive effects had been derived. Arguably, a distinguishing stylistic feature of the digital era is the long, uninterrupted single shot—and even the creation of the single-shot film, like *Russian Ark*—which is far more easily achieved now and may well have become more common. There are a number of notable instances in recent "digital" films of very long, complex moving or tracking shots, lasting many minutes, longer than most (if not all) single-shot sequences from the "film" era.

Perhaps the most notable and familiar instances are in *Birdman* (2014) and *The Revenant* (2015), both directed by Alejandro G. Iñárritu, who has been one of the most prominent filmmakers emphasizing the new possibilities of the long take in the era of digital editing, of digital montage. There are many other recent films, though, that take advantage of the new digital technology to create long, complex sequence shots, some of which are indeed true single takes, but many of which consist of several separate takes that have been invisibly combined, seamlessly blended. Audacious, even ostentatious, examples can be found in otherwise very different films, from *Children of Men* (Alfonso Cuarón, 2006), *Atonement* (Joe Wright, 2007) and *The Secret in Their Eyes* (Juan José Campanella, 2009), to *The Avengers* (Joss Whedon, 2012), *Gravity* (Cuarón, 2013), and the "John Wick" film series (2014–2019). The television show *True Detective*, in episode four of the first season (2014), includes a shot that lasts almost six minutes. The director of the episode, Cary Fukunaga, has said that the shot was created in a single take, as a long, complex tracking shot, without the aid of digital editing or synthesis, and that it was only with the fourth take that they finally succeeded.[78] It is significant, though, that such declarations are thought necessary now, as it is effectively impossible to tell if a long shot is the

78 Paul MacInnes, "How We Got the Shot: Cary Fukunaga on *True Detective*'s Tracking Shot," *The Guardian* (March 17, 2014). https://www.theguardian.com/tv-and-radio/tvandradioblog/2014/mar/17/true-detective-cary-fukunaga-tracking-shot.

effect only of elaborate camera movement, as in *True Detective*, and on the more epic scale of *Russian Ark*, or if it is the result of editing that can now so easily be obscured.

In any case, the long shot or long take has seen a kind of resurgence as a technique, whether in a genuinely single-take form or as a digital composite. In her analysis of *Gravity*, Lisa Purse has described what she calls the "return to the mobile long take," which she presents as a distinctive feature in the era of digital editing.[79] Purse presents several factors that have led to the return to the specific technique. "Computer-generated images, digital compositing, and the move from time-limited storage capacities of tape and celluloid to digital capture have provided the technological conditions for a return to the mobile long-take in film."[80] The phrase "digital compositing" is often used to describe new digital editing, that often does not look like editing, and indeed produces effects, like the long take, long moving or tracking shots, traditionally thought to be an alternative to editing. A distinct stylistic effect is created, a unique feeling, achieved either with cameras that are no longer hindered by the problem of limited storage, by the length of film that can be fit into them, but that are now liberated, unconstrained by the traditional material restrictions, able to record for as long as the footage can be "inputted" with computer processors on ever larger-capacity hard disk storage systems, or with digital editing systems that can seamlessly blend takes together. In a certain basic respect, of course, while such long takes are often created through digital compositing, these are essentially no different than the illusion of spatial and temporal integrity created with disguised cuts in a film like Alfred Hitchcock's *Rope* (1948). Yet, whether synthetic compositions or actual single-take shots, there seems to be a renewed interest in a non-montage aesthetic in the creation of sequences, if not whole films, that are not obviously "edited," that consist, or seem to consist, entirely or mainly of long takes and complex camera movements, constructed "invisibly," rather than "visibly" edited.

Indeed, there is a growing number of films that consist *only* of a single shot, a genuine single take, the most significant, perhaps, being Sokurov's *Russian Ark*, the central example offered by Rodowick as an instance of the new phenomenon of the "digital event," which Rodowick contrasts with traditional editing or montage. The film consists of one, complex,

79 Lisa Purse, "Working Space: *Gravity* and the Digital Long Take," in *The Long Take: Critical Approaches*, eds. John Gibs and Douglas Pye (London: Palgrave Macmillan, 2017), 221.
80 Ibid.

sweeping camera movement, taking the viewer through the corridors of the Winter Palace, now part of the Hermitage Museum, in St. Petersburg, the formal gesture employed to recount the history of Russia since the eighteenth century. It is in fact a single point-of-view shot, presented through the visual perspective of a narrator, whose voice we hear, but whom we never see, as he travels through the centuries, represented by his passage through the palace as he moves from room to room. The film was shot on uncompressed, high-definition video, with a newly available digital video camera, the Sony HDW-F900, stored directly on to a hard disk, rather than to tape, which would have required compression and some loss of quality. The disk was apparently capable of storing almost two hours of uncompressed video—not infinite, by any means, but enough to capture, in a single take, material sufficient for a feature-length film—and in the end the shot took just over ninety minutes to complete. (The film is ninety-six minutes long, with credit sequences added.) The irony of Sokurov's decision *not* to edit was not lost on the many critics who observed at the time of its release that the end of the era of montage may well have been initiated by a Russian filmmaker, in a gesture that could be interpreted as an explicit repudiation of the most influential tradition of the formal use of editing, namely the Soviet montage style, especially as championed by its most fervent advocate, Sergei Eisenstein.

Sokurov's explicit refusal to edit can be understood as a kind of refusal of the montage tradition in Russian cinema, and of the revolutionary politics it was allied with. As Dragan Kujundzic observes, though, the matter is not that simple. While consisting of a single shot, Kujundzic describes the complex *mise-en-scène*, the various contrasts, juxtapositions, and coincidences in the film, as historical characters come and go, as transitions are created by passages through arches, corridors, and doorways, and especially as a kind of collage of images is created, as we see the many paintings, sculptures, and other works in the museum. "The movie," he says, "in that respect appears as an amazing achievement, succeeding in pulling off a colossal 'montage of attraction' and choreography."[81] Still, for Kujundzic, it is more than a mere revision of the concept of "montage," which may now be seen to be effected more through camera movement and *mise-en-scène*, but a refusal of the origins of montage in the specific editing styles of the early Soviet filmmakers, and of the era of revolutionary politics with which it is associated. At the center of the film, he says, is the "erasure of the entire

81 Dragan Kujundzic, "After 'After': The *Ark*ive Fever of Alexander Sokurov," *Quarterly Review of Film and Video* 21, no. 3 (2004): 222.

Soviet period of which in fact nothing is shown, except its absence."[82] Moreover, he says, "the film enacts the erasure of the dominant cinematic tradition, the 'original' montage of attraction by Sergei Eisenstein, and filming life as it is with the numerous cuts attempted by the *kino-eye* of Dziga Vertov."[83]

This is achieved mainly through Sokurov's choice of the Hermitage Museum, the old Winter Palace, as the setting for his film, and a central location in Eisenstein's *October* (1927) and in Pudovkin's *The End of St. Petersburg* (1927), as Kujundzic notes. In fact, and as Kriss Ravetto-Biagioli has observed, the reference especially to *October* is quite explicit. She notes that, in Sokurov's film, he "reverses the frenzied finale of *October* that depicts the Bolsheviks storming the Winter Palace, climbing the famous Jordan staircase and unseating Kerenski and the provisional government, which historically took place in the very same room we see in *Russian Ark*."[84] Ravetto-Biagioli does not offer a detailed comparison, but looking carefully at the two films the similarity in composition is striking. For a moment in *Russian Ark*, we are presented with the almost identical perspective represented in *October* (figs. 5–6). As the scene proceeds in Sokurov's film, we move through the colonnaded spaces of the Palace that are the same ones shown in the concluding sequence of *The End of St. Petersburg*. Rather than a throng of uniformed and gowned aristocrats, though, who are descending the staircase in *Russian Ark*, Pudovkin shows a charwoman climbing them, now empty after the assault on the Palace (figs. 7–8). Ravetto-Biagioli notes significantly that, while Sokurov represents the last two hundred years of Russian history, with the single sweeping camera movement through the Hermitage Museum, with characters from various significant moments in that history entering and exiting the frame, the Bolshevik Revolution of 1917, and the entire Soviet era, is elided, "left unnamed and unrepresented in the film."[85] The only reference to it is in the subtle but significant formal allusions to these classics of Soviet montage cinema—which, I would suggest, are indeed significant representations in the film of the historical fact of the Revolution.

Sokurov was quite clearly making a point, implicitly rejecting the montage aesthetic, or at least a certain specific conception of it, bound

82 Ibid., 227.

83 Ibid.

84 Kriss Ravetto-Biagioli, "Floating on the Borders of Europe: Sokurov's *Russian Ark*," *Film Quarterly* 59, no. 1 (Fall 2005): 19–20.

85 Ibid., 19.

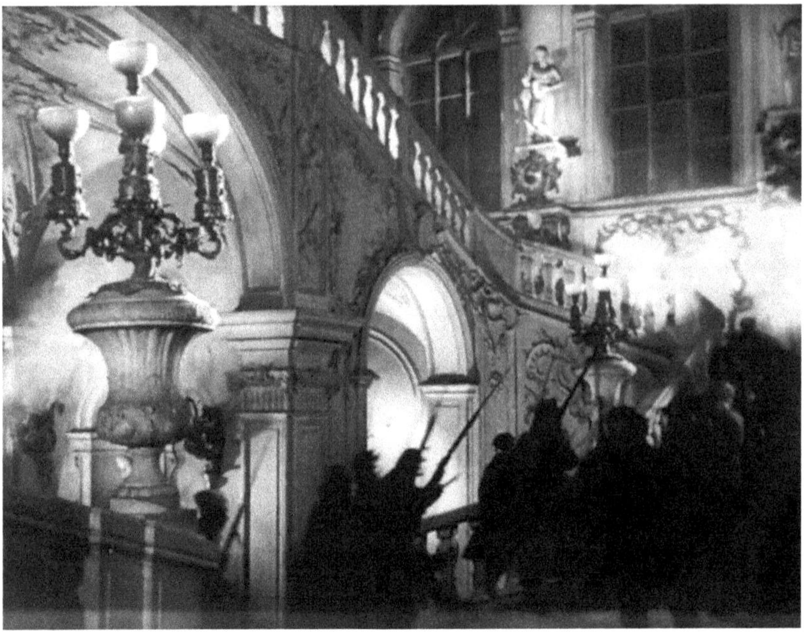

Fig. 5 The stairs in *October* (Sergei Eisenstein, 1927).

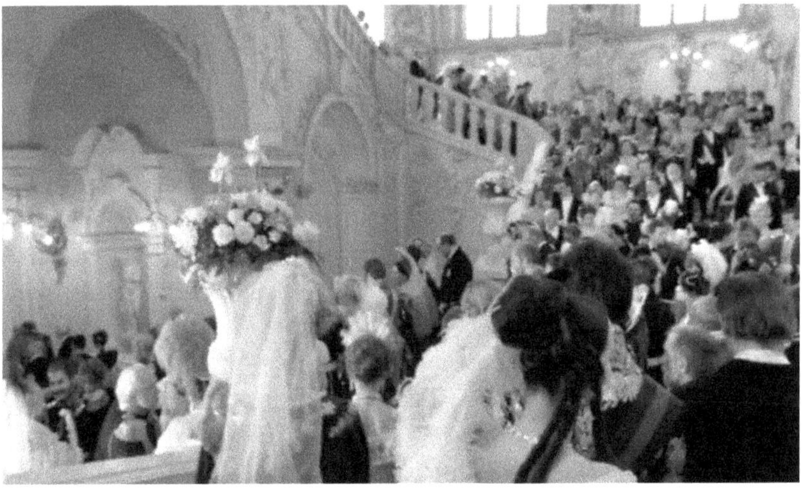

Fig. 6 The stairs in *The Russian Ark* (Alexander Sokurov, 2002).

up as it is with the revolutionary politics of Bolshevism. It is the nature of the rejection that is debated. Rodowick argues that, despite the fact that there are no cuts, that the film was shot in a single take, it must be properly analyzed as an *edited* work—the very nature of the editing transformed, though, by digital technology, and thereby diminished. "*Russian Ark* is

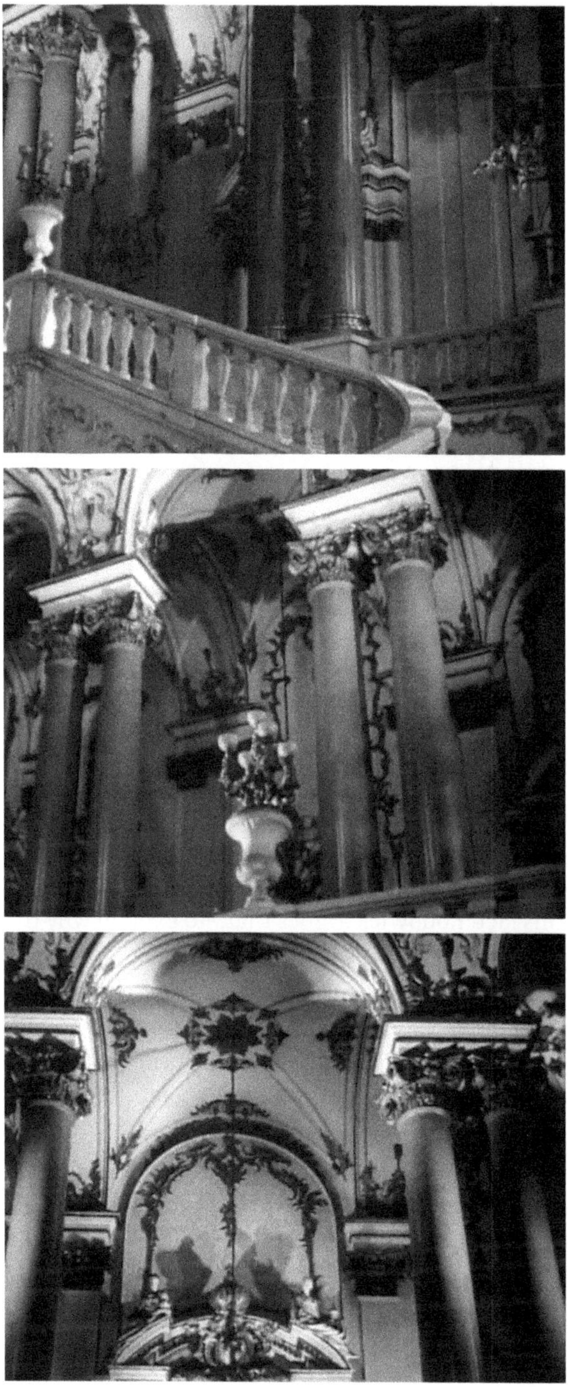

Fig. 7 [a, b and c] *The End of St. Petersburg* (Vsevolod Pudovkin, 1927).

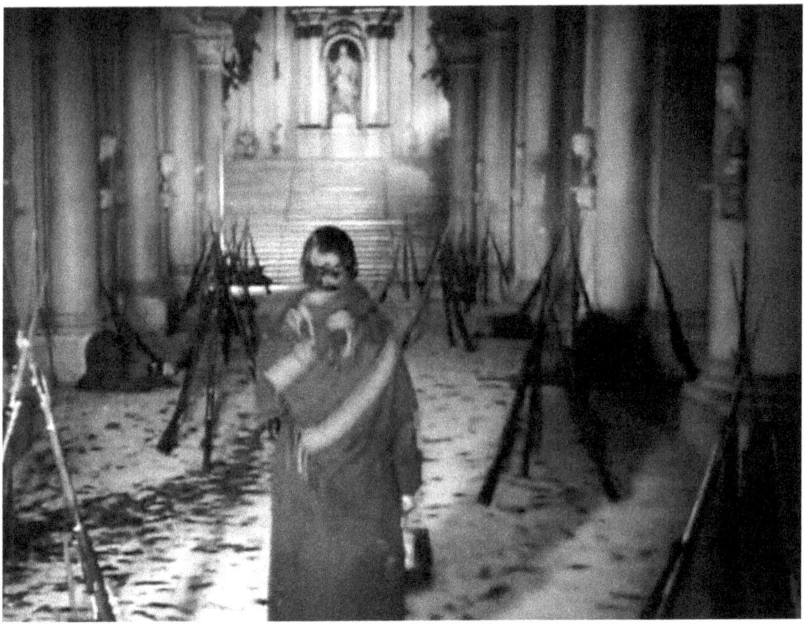

Fig. 8 *The End of St. Petersburg* (Vsevolod Pudovkin, 1927).

a work captured in one go, as it were," he notes. Nevertheless, he insists that it is "a *montage* work, no less complex in this respect than Sergei Eisenstein's 1927 film *October*."[86] For Rodowick, though, and in this very respect, it is a "failed film," largely because it does not acknowledge the fact that it is a work of montage, of *digital* montage, manipulated and altered after the fact with digital editing technology—indeed it seems to deliberately suppress this fact. Rodowick argues that, "to explain why the movie cannot be considered one long take or a single shot goes a long way toward explaining how digital cinema transforms both these concepts."[87] In its form as digital information, Sokurov's film was indeed subjected to considerable post-production processing, so that, for example, lighting elements are imported, colors changed, items of the *mise-en-scène* added and removed, and details like falling snow incorporated into scenes—"even a perspective algorithm has been invoked," Rodowick notes, "to change the relatively normal view to wide angle, thus distorting space expressionistically."[88] For Rodowick, "the style of most digital compositing is to suppress the difference in the apparent spatial unity of the constructed

86 Rodowick, *Virtual Life of Film*, 165.
87 Ibid.
88 Ibid., 167.

image," arguing that such alteration in *Russian Ark* "illustrates well how the separability of image components into discrete layers brings something fundamentally new to digital images: the possibility of what Manovich calls 'spatial' montage."[89]

Whether the alteration of elements within the frame is indeed best described as a kind of "montage," as a new kind of *editing* within the space of the shot itself, it is certainly the case that new tools are indeed available to filmmakers, and that the manipulation of elements of the shot is easier now, arguably a part of a more comprehensive process of editing, which may well have to be reconceived as a technique. There are indeed obvious examples of stylistic experimentation, many recent films that seem like a direct response to, a deliberate use of, new digital technologies, that emphasize the apparent increase in control, the expansion and extension of the powers to alter and manipulate, that is offered by the technology. Examples, that is, of works that are knowingly and deliberately pursing the effects of such new technology can readily be found. While perhaps historically and geographically significant, Sokurov's film was not the first to take full advantage of the new possibilities that new cameras and editing systems offered the filmmaker seeking to compose in a long, single shot. In 2000, Mike Figgis made *Timecode*, which in fact consists of four separate single-shot long takes, which are shown simultaneously in equal quadrants of a split-screen (fig. 9).

The arrangement, it could be argued, is a visual manifestation of the kind of control over the image itself that Figgis now has, as a result of the very technique of "timecoding," which was an important first step in the development of digital editing systems. The title of the film refers to the technical innovation in video recording that made electronic, and then digital, editing possible. Editing, of course, requires some way of identifying, and keeping track of, the various elements, the pieces of film or the segments of video, that are being assembled. *Film* had numbers and codes—"keycodes"—printed on the edge of the strip, allowing editors to know what part of the original negative corresponded to the work print they were cutting, and to keep track generally of their footage. *Video* had no such physical identifying markers, until 1967, when, as Thomas Ohanian explains, in one of the first handbooks for the new digital editing systems, "a method of consistently identifying videotape frames was developed. ... Timecode, in the form of hours, minutes, seconds, and frames, such as 01:05:12:20, is a signal that is recorded onto videotape and that identifies

89 Ibid., 168–69.

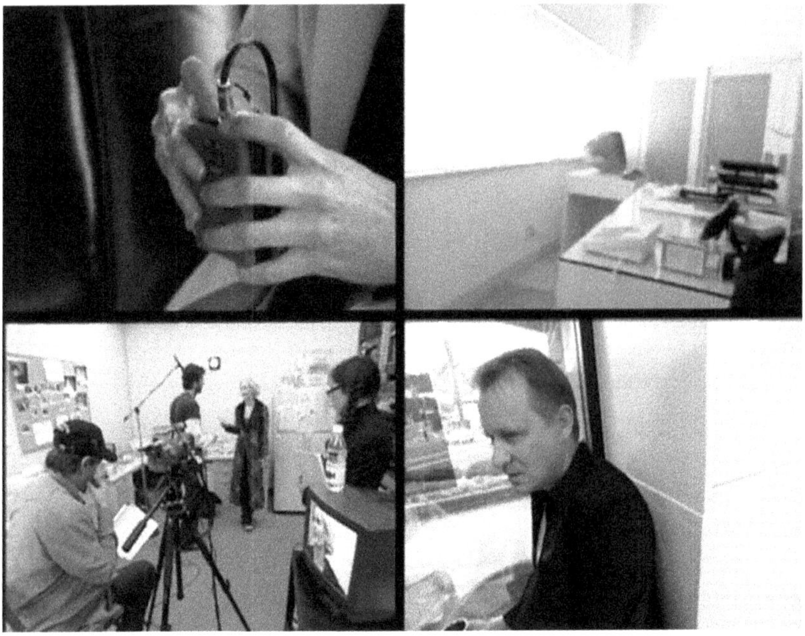

Fig. 9 *Timecode* (Mike Figgis, 2000).

each and every video frame."[90] This was, he says, "revolutionary." The numbers could be easily "read" by computers, which allowed for almost immediate access to any particular frame, ultimately making genuinely random-access, non-linear video editing possible. The "digital era" in editing begins, one could argue, with "timecoded" videotape, and is fully realized with the introduction of comprehensive computing systems that convert the contents of video footage into machine-readable numerical values, essentially applying the logic of the timecode, which had initially identified only large sections of video recordings, to all of the individual elements of the recorded information on the smallest scale possible. Every bit of the representation, so to speak, would thus become available for manipulation, for a comprehensive mode of editing, as the basis for a new aesthetics of digital montage. *Timecode* (the film) revealed the possibilities that the "timecode" created, but which perversely, at the same time, possibly heralded the end of "editing" understood in any traditional sense, that is, the breaking up of the action, and the rearrangement of narrative elements, which had become no longer necessary in strictly technical terms, as a solution to the

90 Thomas Ohanian, *Digital Nonlinear Editing: New Approaches to Editing Film and Video* (Boston and London: Focal Press, 1993), 16.

particular material problem of the length of the filmstrip. The aesthetic of *Timecode* subtly represents the effective atomization of the film image, the breaking up of the image and the sound into manipulable elements, while, at the same time, revealing the possibilities of the long take, the technique that had long been understood to represent the integrity of the shot, the spatio-temporal inviolability of the motion picture image. The very aesthetic complexity of the new formal possibilities created by digital editing systems is the basis of the design of Figgis' film.

As far as I know, no similar experiment in simultaneous single-shot sequences on a split-screen has been undertaken since, at least not on the scale of Figgis' film, but there is, as Lisa Purse notes, an unmistakable return of a long-take aesthetic, the somewhat paradoxical result of an extension of the powers of editing. There have indeed been several significant single-frame, single-shot films made recently, in the era of digital editing, that depict an entire sequence of events *without* an edit, without any obvious or apparent cuts. They are constructed, like *Russian Ark*, as true *single-take* films, composed (it is claimed) without any invisible digital synthesis, without the aid of new digital editing systems. They may well, as Rodowick notes in the case of *Russian Ark*, have been subjected to other kinds of post-production finishing or processing, which the new systems enable or allow, but they are created initially, at the moment of production, by simply turning the camera on and letting it run, moving it as necessary, often in quite elaborate and complex patterns, until the event is completed, until the story is told—in, that is, a single take, presented as a single shot. Notable examples include *PVC-1* (Spiros Stathoulopoulos; Colombia, 2007), *Fish & Cat* (Shahram Mokri; Iran 2013), *Ana Arabia* (Amos Gitaï; Israel, 2013), *Agadam* (Mohamad Issack; India, 2104), and *Victoria* (Sebastian Schipper; Germany, 2015). Each of these is advertised as having been completed in a single take, without any editing, digital or otherwise. A perhaps extreme example is *Lost in London* (2017), starring and directed by Woody Harrelson (in his directorial debut), which extends the idea in such a way that suggests a breakdown between the very idea of film as constructed audio-visual artifact, to be viewed after the fact of production, and bearing the *marks* of production most significantly in the form of visible editing. The film is touted as the first to be *broadcast live* into movie theaters for its premiere, *as it was being filmed or recorded*, which required that the single, almost one-hundred-minute shot, constructed as a single take, be successfully completed the first time. (It was, reports say.) Sokurov, by comparison, is widely known to have succeeded in *Russian Ark* only on the fifth attempt—such are the challenges of single-shot filmmaking. Such films raise basic aesthetic questions about the art of filmmaking, and about the artistic possibilities now afforded by new digital editing systems, but as aspects of the always changing contexts within which artists work, rather than as a challenge or threat to the cinema's ontology as an art. The decision not to edit is comprehensible only in relation to the idea of editing as an artistic choice.

***

Such single-shot films can only be understood in light of the fact of editing, and on the basis of a clear account of editing as a technical and aesthetic phenomenon, which a filmmaker may choose to use *or not*. Explaining editing, though, much less developing an explicit theory of the technique, has been something of a challenge, insofar as it has become so bound up with supposedly ontological issues. Lucy Fischer notes that editing has long been a central topic for film theorists, understood precisely as a particularly significant and perhaps defining technique, as "a major component of cinematic form. As such," she says, "it has received considerable attention in the annals of film theory. Yet it is not a simple term to define."[91] She summarizes the various attempts, which tend to explain editing by comparison—in relation to realism, to creativity, and to the other arts; as "mimicking" the mind, the psyche, or as reproducing the process of cognition; and finally, as a mode of communication, a kind of "language," as the means, in a word, for expression—all versions of what I am describing as "ontological" analysis. Even if reduced to this last specific function, though, as it very often is, understood as the primary expressive technique in the cinema, editing still proves to be difficult to explain. As Valerie Orpen has noted, reviewing the literature on the subject, "editing as an expressive technique is taken for granted. We all know that it is expressive, but it is more difficult, uncomfortable even, to explain why and how."[92]

Editing has long been thought to be the primary source of significance and meaning in the cinema, the technique that has provided the cinema with its expressive capacities. Most accounts of editing begin dutifully with the famous observation by V. I. Pudovkin, that the "foundation of film art is *editing*."[93] To establish his claim, he argues that "a film is not

91 Lucy Fischer, "Film Editing," in *A Companion to Film Theory*, eds. Toby Miller and Robert Stam (Malden, MA and Oxford: Blackwell, 1999), 64.

92 Valerie Orpen, *Film Editing: The Art of the Expressive* (London: Wallflower Press, 2003), 3.

93 V. I. Pudovkin, *Film Technique and Film Acting*, trans. Ivor Montagu (London: Vision Press, 1954), xiii; emphasis in original.

shot, but *built*, built up from the separate strips of celluloid that are its raw material."[94] He compares the process to writing, to the use of language. "If a writer requires a word," he says, "the single word is only a raw skeleton of meaning. ... Only in conjunction with other words, set in the frame of a complex form, does art endow it with life and reality."[95] It is in this respect, he insists, "that editing is the creative force of filmic reality, and that nature only provides the raw material."[96] Pudovkin describe cinematic expression precisely as "filmic speech," the "language of separate images, separate elements, shots, that follow one another in a certain sequence."[97] Others have made similar claims. If the analogy with speech, with language, is not always as explicit, there is a sense that *meaning* in the cinema is the effect primarily if not exclusively of editing. Rudolf Arnheim distinguishes the techniques of editing from the other formal elements of the cinema, elevating it to the highest rank, emphasizing its value as the main or even the single source of meaning in film, or at least as the technique for fulfilling the meaning-making function of cinema. "In montage," he says, "the film artist has a first-class formative instrument, which helps him to emphasize and give greater significance to the actual events that he portrays."[98] Even Siegfried Kracauer is willing to admit the central importance of the technique, specifically emphasizing its expressive power. As one of what he describes as the "technical properties of film," he says, "the most general and indispensable is editing. It serves to establish a *meaningful* continuity of shots."[99] Kracauer emphasizes that his "interest lies not with editing in itself ... but with editing as a means of implementing—or defying, which amounts to the same thing—such potentialities of the medium as are in accordance with its substantive characteristics."[100] Implied here, though, is the sense of editing as the mechanism needed—Kracauer indeed calls it "indispensable"—to reveal or to emphasize the meaning already in the shot, to create a "meaningful continuity of shots." It is important for Kracauer, though, from his "realist" perspective, that editing not limit the inherent

94 Ibid., emphases in original.
95 Ibid.
96 Ibid.
97 Ibid., 93.
98 Rudolf Arnheim, *Film as Art* (Berkeley: University of California Press, 1957), 89.
99 Siegfried Kracauer, *Theory of Film: The Redemption of Physical Reality* (Princeton: Princeton University Press, 1960), 29; my emphasis. The "technical properties" of film, of course, which also include, for Kracauer, special effects, lighting, and so on, are contrasted with what he calls the "basic properties," which are primary, he says, insofar as they are "identical with the properties of photography" (29).
100 Ibid.

indeterminacy of any individual shot, even as he admits that shots must be arranged in relation to other shots, such arrangement necessarily imposing a potentially singular meaning. He proposes a "basic editing principle," which is that "any film narrative should be edited in such a manner that it does not simply confine itself to implementing the intrigue but also turns away from it toward the objects represented so that they may appear in their suggestive indeterminacy."[101]

As opposed to such a "realist" account of editing, the "formalist" film theorists have emphasized the power of editing as the means precisely to fulfil the basic element of the shot, which on its own does not yet possess meaning, or which is indeed indeterminate, or in need of articulation. "The single image," Arnheim says, "arises from a recording process, which is controlled by man, but which, regarded superficially, does no more than reproduce nature."[102] But, he says, "when it comes to montage man takes a hand in the process—time is broken up, things that are disconnected in time and space are joined together. This looks much more like a tangibly creative and formative process."[103] It is easy, he suggests, to see film art itself beginning with the techniques of editing, as the most obviously and manifestly intentional aspect of filmmaking, to identify the origins of the cinema with the invention of editing. André Malraux, for example, famously and influentially declared that the art of film did not begin with the mere capturing of moving photographic imagery but only with editing. "The problem," he says, "was not one of rendering the movements of a person within a picture, but of conveying tempo, the sequence of successive moments. It was not to be solved mechanically by tinkering with the camera, but artistically, by the invention of 'cutting.'"[104]

101 Ibid., 71.
102 Arnheim, *Film as Art*, 87–88.
103 Ibid., 88.
104 André Malraux, "Sketch for a Psychology of the Moving Pictures," in *Reflections on Art: A Source Book of Writings by Artists, Critics, and Philosophers*, ed. Suzanne K. Langer (Baltimore: The Johns Hopkins Press, 1958), 320; emphasis in original. Translated as "cutting," the term in French that Malraux uses is "*découpage*." This is usually understood as the preparatory work, anticipating the editing to follow, by carefully arranging the elements of the shot to function according to the plans for editing. The practice of decoupage is understood as an important aspect of the transformation of mere editing to montage. See André Malraux, "Esquisse d'une psychologie du cinéma," *Verve* 2, no. 8 (June 1, 1940), 69-73. Originally published in *Verve*, the essay was revised several times, and published elsewhere. For a comparison of the versions in French, see François Albera et Jean-Paul Morel, "Esquisse d'une psychologie du cinéma," *1895. Mille huit cent quatre-vingt-quinze* 76 (2015). Accessed July 12, 2024. http://journals.openedition.org/1895/5020 ; DOI : https://doi.org/10.4000/1895.5020

It is usually as an *expressive* technique, as such a creative or "formative" process, that editing is conceived, beyond the mere fact of recording, so that shots are understood as the basic units of meaning, awaiting, though, the proper combination that will grant them a specific intended meaning, as an "expression." It is this kind of *syntactical* power that, for example, Jean Mitry emphasizes. "Aesthetically speaking," writes Mitry, "editing (or *montage*) is not so much the effect of matching images according to a logical continuity as of giving the images a meaning beyond that of the information presented in them, creating a new power deriving from the juxtaposition of two or more shots *which then assume a value they could never have outside that association.*"[105] To emphasize the expressive or communicative power of editing, Mitry offers a familiar analogy. He argues that, through montage, "*shots in a sequence behave in the same way as words in a sentence where subject, verb, and object only begin to have meaning when they are related to each other.*"[106]

Even if such an analogy with language, the equation of shots with words, and of sequences with sentences, is rejected, there is a common tendency to think of editing in explicitly expressive or communicative terms. Noël Carroll is one of the few to have actually worked this out theoretically. "To understand editing," Carroll has said, "we must understand it as a form of communication."[107] He resists the sort of linguistic models that were common at the time of his writing, though, which compared the cinema to language, understood as the paradigmatic mode of expression or communication. This was the approach of the emerging methods of structuralism and semiotics, but which he traces back to such figures as Eisenstein, Pudovkin and the other Soviet montage theorists. In contrast, Carroll offers a cognitivist

105 Jean Mitry, *The Aesthetics and Psychology of the Cinema*, trans. Christopher King (Bloomington and Indianapolis: Indiana University Press, 1997), 68; emphases in original. It is worth noting that, in the original French, Mitry does not make the distinction made in the English translation, between "editing (or *montage*)." Mitry simply says, "au point de vue esthétique, le montage est bien moins le fait de raccorder les images selon la continuité logique de l'action que de leur donner un sens qui dépasse celui du donné représenté." The word "*montage*" is the most commonly used in French to mean "editing." See, Jean Mitry, *Esthétique et psychologie du cinéma I: Les Structures* (Paris: Éditions Universitaires, 1963), 161. Mitry, as I have noted earlier, in Chapter 1, distinguished between what we call "editing" in English, what he often just calls "*montage*," in French, but which he also calls "*montage conditionnel*," as a strictly technical aspect of filmmaking, and the *effect* of editing, "*l'effet-montage*," or what we call in English simply "montage." The distinction is not always clearly maintained in the English translation.

106 Mitry, *Aesthetics and Psychology of Cinema*, 161; emphasis in original.

107 Noël Carroll, *Theorizing the Moving Image* (Cambridge: Cambridge University Press, 1996), 403. From a reprinted essay originally published as Noël Carroll, "Toward a Theory of Film Editing," *Millennium Film Journal*, 3 (Winter–Spring, 1979): 80.

theory of film editing, arguing that "editing is a communication practice based on cutting which prompts the spectator to infer the significance of a shot chain in terms of the best available account of the shot chain."[108] While rejecting any simple film/language analogy, he nevertheless insists that editing is the main source of significance in the cinema, of an expressive power, insofar as it is the basis for cinematic communication, that editing is a "communication practice that functions by eliciting inferences from spectators."[109] Later, though, Carroll would admit that his attempt to develop a basic theory of film editing was a failure, that the problem was much harder than he had originally thought. In what he describes as his "earliest published attempt at film theory," his account of editing had what he says was a "glaring problem," which is that "its central terms of analysis—inference and interpretation—are simply too vague."[110] To say that editing is the basis of expression in the cinema, of communication, was much more difficult to explain and substantiate than Carroll had expected.

Having tried earlier, and, on his own account, failed to elaborate a theory of editing, Carroll has subsequently sketched a broader and more expansive outline of what he describes as a theory of cinematic sequencing, of the compositional logic for the ordering and arrangement of elements in film, *with or without* actual editing. Editing, of course, is the most *common* means. As Carroll says, "most motion pictures are comprised of more than one shot. Typically, shots are combined into sequences. They are added together in order to perform certain functions. ... Cinematic sequences," he says, quite simply, "are typically assembled by means of film editing."[111] A common pattern in film narrative, for example, is to shift from close-up shots, providing information about specific details, to long shots, which allow for a synoptic perspective, to see broader contexts. "In editing," says Carroll, "the space between the close-up camera position and the long view is excised. Hence, the idea of a 'cut': something has been *cut* out."[112] Cutting out material is a powerful means for constructing a story in film, or any other kind of expression, and editing has become more or less indispensable, if not absolutely necessary, insofar as it can effectively and efficiently perform this function. It has always, of course, been possible, as Carroll notes, to

108 Carroll, *Theorizing the Moving Image*, 417–18.
109 Ibid., 416.
110 Noël Carroll, in the introduction to the section "False Starts," in *Theorizing the Moving Image*, 374.
111 Noël Carroll, *The Philosophy of Motion Pictures* (Malden, MA and Oxford: Blackwell, 2008), 116.
112 Ibid., 117.

construct a sequence without editing. A sequence, that is, "can also be built up within a single-shot via camera movement—the image can go from the close-up to the long shot by moving or tracking the camera backwards, thus achieving an effect that is functionally equivalent (in terms of information conveyed) to cutting between two different camera positions."[113] Rather than cutting something out, it can instead be left in. As he says: "In camera movements ... the 'space between' is exhibited."[114]

There may be more emphasis in the digital era on what we could call, following Carroll, an "exhibitive" style. What I want to stress here, though, is the phrase "functionally equivalent." Eisenstein, by the time he came to write the essay "Montage 1938," had implicitly accepted the sort of functionalist analysis that Carroll develops, acknowledging that "montage," as a general principle of cinematic sequencing, could be achieved *either* by editing, or by *other means*, such as camera movement. He had, that is, transformed his view so that the value of a stylistic technique was not reducible to any particular material or physical fact about the medium, to any specific limitation or technical constraint. Montage was no longer understood to derive from the necessity of editing—and it does not lose its value in the digital era, now that editing is, strictly speaking, no longer necessary. In this respect, and as I have been arguing, the familiar debates about the relative value of montage versus the long take, the Einsteinian versus the Bazinian position, can be resolved, as the question becomes "functional" rather than "ontological," as this distinction is made by Genette, with whose argument I began.

A convincing theory of editing as communication or as expression has yet to be elaborated. Indeed, few have deigned to offer an explicit theory of editing on such grounds, an account that could in fact substantiate the claims, effectively just assertions, of the kind that I have just surveyed: that editing is like language, or that it is simply the arrangement of shots in a particular (presumably the proper) order that turns those individual shots into some larger, meaningful whole, according to a syntactical logic that makes expression and communication possible in language. It was, of course, and perhaps somewhat ironically, Christian Metz who revealed the flaw in such thinking, who demonstrated, with the tools and concepts of linguistics itself, that shots were not at all like words, that editing did not proceed according to any truly syntactical logic, like that which governs the ordering of basic elements, of letters, or "phonemes," into words, and

113 Ibid., 116.
114 Ibid., 117.

words into sentences. Such a view, he says, originates in Soviet montage theory, in the first and most influential theories of editing, with Pudovkin, and with Eisenstein. Using the basic terms of Saussure's general linguistics, Metz argued that there may well be a "language" of film, "*un langage*," but that there is no underlying language "system," what Saussure called *langue*.[115] Metz sought to preserve a structuralist, semiological account of cinema, that would not, though, be reducible to the simplification that editing patterns bear any direct resemblance to syntactical structures. Yet editing still needed an explanation. In a later work, in which he shifts to the broad concept of "enunciation," derived primarily from the work of Emile Benveniste, Metz is still concerned with the problem, noting that "the considerable weight of elements and arrangements that film mixes into or adds to simple representation, of which *montage*, for the majority of theorists, is the symbolically totalizing expression."[116]

Montage, that is, is editing understood precisely as an expressive system, and one that would (should) presumably function according to a logic of syntax. Understood in the simplest sense, of course, film editing is indeed merely the sequential arrangement of shots, separated by a cut or by some other transition like a dissolve. The new tools of digital editing do not change this, even as the tools themselves have changed completely. While the "cut" was originally understood literally, the result of actually cutting a strip of film into a smaller piece and connecting it to another piece, cuts are now made "virtually," on the computer, rather than with a pair of scissors or a blade. This is a significant and comprehensive transformation of the basic technology of editing, technology that had, until about the early 1990s, remained more or less the same since editing emerged as a technique. Still, despite the significance of this technical change, the aesthetic effects of editing are the same. One shot is followed by another, one sequence of shots followed by another sequence, building, in necessarily linear fashion, a larger complex whole. Nothing about this basic fact of editing has changed, this fact that has been basis of most attempts to elaborate a theory of editing.

Perhaps something else has changed, though. It is sometimes claimed that the new digital editing systems have increased the number of cuts, that the pace or the rate of editing has quickened, presumably because it is *easier* now to cut, because it does not involve the actual physical efforts of

115 See Christian Metz, "The Cinema: Language or Language System?" *Film Language: A Semiotics of the Cinema*, trans. Michael Taylor (New York: Oxford University Press, 1974), 31–91.

116 Christian Metz, *Impersonal Enunciation, or The Place of Film*, trans. Cormac Deane (New York: Columbia University Press, 2016), 149. The book was originally published in French in 1991.

cutting a material strip of film and gluing or taping it to another piece. This seems plausible, and there is some evidence that the "average shot length" (or ASL) has indeed become shorter, that there is more cutting, that there is, as David Bordwell has argued, a more "intense" form or rate of editing, especially in mainstream Hollywood, or Hollywood-style cinema, possibly as a direct effect of the shift to computer editing.[117] At the same time, though, and as I have considered above, new digital editing technology seems to have reduced the number of cuts, or allowed filmmakers to more easily disguise the cut, creating longer shots, or the impression of longer shots, that are in fact made up of several takes, invisibly combined. Digital editing makes this much easier to achieve, providing the means to effectively disguise cuts, to render them invisible, to make them look like fluid camera movements. Many recent films, edited on new digital systems, feature long, virtuosic shots, often complex moving or tracking shots, that are in fact a kind of digital blend, several shots (or takes) combined and made to look like one.

On the whole, though, there is nothing new or stylistically distinctive about either of these tendencies. There are many examples, throughout the history of film, of fast cutting and just as many examples of slow cutting. One can find short sequences that consist of many shots, and many long sequence shots, moving-camera sequences, that consist of a single shot. While it may arguably be *easier* now to either cut or not cut, such aesthetic choices have existed for as long as there has been cutting, since what we recognize as editing was invented. Moreover, it is very difficult to make any kind of generalizations along these lines, even on the basis of the kind of statistical analyses that have been tried. While it may be the case that the long take has "returned," it would be difficult to determine if in this moment of return there are more long takes than at any other moment in the history of the cinema. Relative rates of cutting are even more difficult to determine. As James Cutting et al. have noted, there is no simple way of measuring ASL across a large sample of films. Sitting down and simply counting cuts is laborious and unreliable.[118] While there

117 David Bordwell, "Intensified Continuity Visual Style in Contemporary American Film," *Film Quarterly* 55, no. 3 (Spring 2002): 16–28. Bordwell describes some specific consequences of the shift to digital editing systems. "Cutting very brief shots on celluloid," he says, "is labor-intensive and complicated, since trims only a few frames long can easily go astray. By cutting on computer, filmmakers can easily shave shots frame by frame." Ibid., 23.

118 James E. Cutting, Jordan. E. DeLong, and Christine E. Nothelfer, "Attention and the Evolution of the Hollywood Film," *Psychological Science* 21, no. 3 (2010): 432–39. The counting of cuts, sitting at a computer and manually noting them with a click or a keystroke, as unreliable as it is, is the basis, nevertheless, of the method called "cinemetrics." As described on the website, the system

are software programs for counting cuts, what Cutting et al. describe as "cut-finding algorithms," these tend not to function very well, as they note, given that they "often confuse motion across frames within a shot with spatial discontinuities across shots. They also do poorly with fades, dissolves and wipes."[119] Given these limitations, they devised their own software application, based in MATLAB programming language, traditionally used for mathematical calculations, automating the process for counting cuts in Hollywood films made between 1935 and 2005, their chosen sample. What they found was that "Hollywood film has become increasingly clustered in packets of shots of similar length."[120] There seems to be a movement towards a kind of equilibrium, so that action sequences, notable for faster cutting rates, have slowed somewhat, and dialogue scenes, traditionally slower, have sped up slightly, such that the authors can observe, generally, that "film editors and directors have incrementally increased their control over the visual momentum of their narratives, making the relations among shot lengths more coherent over a 70-year span."[121]

They do not suggest any direct link, though, between this formal "evolution," as they describe it, and any particular technology or change to technology. The trend towards a greater balance between scenes of action and scenes of talking does not seem related at all to the advent of digital video editing systems, which were introduced only in the 1980s, taken up by the film industry through the 1990s, and which have become the norm only in the last two decades. Their concern is to provide some empirical grounds for the claims of cognitive film theory, which posits a relation between the natural perceptual and conceptual apparatus possessed by human beings and the formal structures of films. They argue that the basic human capacity to pay attention is what governs the evolutionary changes in film editing patterns. They conclude that the general trend in such patterns is towards a formal representation of events in film that can be most easily comprehended given the basic cognitive apparatus we possess that allows us to make sense of the world itself. Given our basic cognitive apparatus, that is, and with regular

is "a tool for measuring and visualizing movie data, in order to reveal the characteristics of films and to create visual 'fingerprints' of them. Information such as the editing structure, color, speech or motion are extracted, analyzed and transformed into graphic representations, that can be easily interpreted or compared side by side." See Cinemetrics, "Info." Accessed July 12, 2024. https://cinemetrics.site/?selection=,,,,,#/info.

119 Cutting et al., "Attention and Evolution of Hollywood Film," 433.

120 Ibid., 434.

121 Ibid.

exposure over the history of the cinema, we have become more capable of making sense of more rapid editing, more concise forms of montage, so to speak. Leaving aside the plausibility of any such claim, and assuming that we can even rely on the empirical analysis that they provide, what Cutting et al. describe is a gradual change in editing style—in fact, a slow process of consolidation of the sorts of techniques developed between about 1910 and 1920—that is more or less unaffected by any changes to editing technology itself, and which is explained instead by the honing of human cognitive capacities.

Still, there is the one effect, which I have introduced already, that may be traced to the advent of digital editing systems, a counterintuitive, even perverse effect. During the era of altered and arguably enhanced editing technology there has been the emergence, in a kind of historical irony, of films that eschew editing altogether, pursuing in comprehensive form what Carroll describes as the "exhibitive" mode—not cutting out anything at all. This may have begun with Sokurov's *Russian Ark*, which is, William Johnson says, "the first feature-length movie ever to consist of a single, uninterrupted 90-minute shot."[122] This is, we could say, and as I have suggested earlier, an indirect effect of digital editing, an effect of the more general change that has made digital editing possible. Camera and sound recordings are now typically rendered as digital information, either converted from so-called "analog" form, from traditional photographic imagery and magnetic tape audio recordings, into digital files, or shot originally with a digital camera and sound recorder, saved onto a hard disk, so that the information can be exported directly to a computer for editing and subsequent processing. Such recording technology allows for ever longer takes. Enough information may be stored on new digital cameras so that an entire narrative may be constructed while shooting, created, let's say, *cinematographically*, that is, during production, rather than by cutting and reassembling the recorded film in post-production, as has been typical. A film may now be created in a single take, presented as a single shot, the length of which is limited only by the capacity of the camera, itself a kind of small computer with a hard drive or other electronic storage space. In the past, of course, the length of a shot was limited by the amount of actual film that could fit into a magazine attached to the camera. While some magazines for 35mm motion picture cameras could fit up to 1,000 feet of film, these were quite bulky and not often used. Most film cameras used in professional production held about 400 feet, which

122 William Johnson, "*Russian Ark*," *Film Quarterly* 57, no. 2 (Winter 2003–2004): 48.

would be enough for a shot of about twelve minutes at most.[123] Even with the largest magazines available, though, no single take could exceed about half an hour, and most filmmakers were used to having roughly only ten minutes or so to work with at any time while shooting, and most finished shots were, of course, much shorter.

With the increased recording or storage capacity of digital cameras, as I have said, there have been several notable films made in the last two decades that consist of a single shot, usually, though, constructed from several shots, or takes, digitally blended together. Strictly speaking, that is, they are edited, but the cuts are obscured. There are also films, though, like *Russian Ark*, and the others that I have mentioned, that are true single-take films, recorded directly onto high-capacity storage disks or hard-drives. The ironic effect of all of this is that films, now, in the era of digital editing, need *not* be edited, or may be made in such a way that appear not to be edited. The fact of editing may be entirely effaced, leaving no visual trace. Where there would have been a cut or a dissolve or some other visible transition, the filmmaker can choose to obscure that trace, and make it appear as though there is a temporal and spatial continuity between what are in fact two separate takes. A filmmaker may also choose to forego cutting entirely, as I say, and shoot an entire film in a single take, as Sokurov and others have chosen to do.

A particularly notable example is Sebastian Schipper's 2015 film *Victoria*. With even more storage space and computing power available than when Sokurov made *Russian Ark*, Schipper shot his film in a single take that is over two hours long—the film's entire running length is 138 minutes. Moreover, Schipper shot on location, mainly outside, on the streets of Berlin, with all of the potential interruptions that could have occurred, creating an even greater challenge than Sokurov had working in the tightly controlled space of the Hermitage Museum in Saint Petersburg. Such films are very rare, though. Most single-shot films made in the digital era in fact consist of several takes but are edited so that they *appear* to be a single shot. Sam Mendes' *1917* (2019) is an example, consisting of several takes stitched together digitally to appear to be a single shot. It was not shot all at once but over several days. In fact, Mendes' film arguably consists of two shots, separated by a fade to black when the main character loses consciousness, allowing a

123 For a good description of the basic technology of filmmaking before the digital era, at the height arguably of the so-called "analog" cinema, see Raymond Spottiswoode, *Film and Its Techniques* (Berkeley and Los Angeles: University of California Press, 1969), especially the chapter on "The Camera," and the description there of the "magazine," 50–51.

temporal ellipsis to be incorporated into the narrative—the several seconds it takes to fade from the black, and to get back to the action, comprises several hours of actual story time, from late afternoon, with the last rays of sunshine casting long shadows, when the soldier is knocked out, to early evening, when he revives.

Here, of course, we are confronted by the significant limitations that are imposed as soon as a filmmaker decides to construct a single-shot film, whether by shooting in a single take or by merging or digitally blending separate takes into an apparently single shot. By doing so, one is deliberately sacrificing the various powerful effects that editing makes possible, the most significant perhaps being the ability to shift in time and space. Mendes wanted to tell the story over a time greater than the running time of the film, to establish the familiar sort of distinction between story and plot, but without the use of editing—or at least any obvious editing. To do so, he was obliged to compress time through the technique of the fade-out, which he presumably hoped audiences would not notice by connecting it to the consciousness of a character, or more precisely to the *un*consciousness of a character, who, like us watching the film, also misses out on the few hours that pass. Sokurov confronted a similar formal problem, as he tells the story of Russia from the eighteenth to the twentieth century, but without the usual markers, used in editing, to indicate the passing of time, without, for example, any obvious ellipses, at least not in the familiar form, with a fade or a dissolve or a straight cut. He solves this problem by introducing a supernatural element. We are guided by a character, apparently a ghost, a spirit, who can travel through time, and whom we are able to accompany, as we move with him from the eighteenth century to the present, covering about two hundred years of Russian history while nevertheless confined to, moving only within, the spaces of the galleries and hallways of the museum—itself conceived of as a kind of "time machine," a building, an architectural vessel, within which time itself is indeed condensed and subordinated to the logic of the spatial presentation of art works and artefacts.

Critics like William Johnson have noted the correspondence between Sokurov's formal gesture, his "single-shot strategy," and the film's subject and themes. Sokurov had to abandon the take at least two times (some reports say four times) before he was able to successfully complete it. Given that the disk he was using could not be recorded over, he was losing storage space with each false start. As Johnson notes, describing the technical and formal challenge, the two false starts tell us something about the film's themes, when he says that "the sheer precariousness of the project—could it have survived, financially and psychologically, if there had been a major accident during the third take?—reflect the precariousness of history and of life itself. Ultimately," he says, "*Russian Ark* represents a provisional victory over time."[124]

Schipper, in *Victoria*, similarly uses the "precariousness" of the single-take technique to thematic ends, constructing a narrative that benefits from the spatial and temporal limitations of the single shot, tracing the events that occur in real-time, in the two hours of early dawn, when a group of young people get caught up in an armed robbery. The tension is increased by the fact that we can see each step of the process that leads them, fatefully, to commit the brazen criminal act. Their desperate, ill-fated actions are emphasized effectively by the fact that there is no cutting, so that there is no interruption in the unrelenting pace of events, accentuated even further by the almost constantly moving camera. The familiar story of doomed criminals is given a new formal emphasis with the decision not to edit, so that we as viewers cannot look away. We are swept along just as fatefully as the characters themselves. Similarly, in the aptly titled film *One Shot* (James Nunn, 2021), which also consists of a single shot (but not made in a single take), a basic (if simplistic) relation is established between the efforts of a group of special forces soldiers, who have only "one shot" to kill their target, and the film's formal structure, which is uninterrupted by any (visible) cuts, itself apparently "one shot."

The pun in Nunn's title suggests that this has become a recognizable formal strategy, potentially understood by audiences who would get the joke, understand that the "one shot" the soldiers have is literally represented in the "one shot" comprising the film. Indeed, it may well be that there are now more single-shot films, made in the last twenty or twenty-five years (although I have not, I will admit, actually counted), than those made in the entire features era, from, that is, about 1915 to 2000. There are certainly very many films now that are either truly single-shot films, or that, like *1917*, have only one or two visible cuts or transitions, and that consist of multiple takes blended into very long shots. This can only really be explained by the fact of the advent of digital editing systems, and by the development of digital video cameras. In this respect, the aesthetics of digital montage could be reduced simply to the subordination of the values of editing—understood traditionally as the cutting of physical strips of film, and of all the productive limitations and constraints that that imposed—to the values of digital manipulation, which includes most tellingly the potential erasure of the cut, and the effective subordination thereby of the shot, long understood

124 Johnson, "*Russian Ark*," 51.

as the most basic aesthetic unit of film. The tension, that is, seems to be between the strict limitations that the shot imposed on artistic intention and the radical aesthetic and expressive freedom provided by digital editing.

Such films, or such sequences within films, though, it could be argued, are made possible by, or encouraged as an already existing possibility that is made easier by, the technical affordances of digital technologies—with lightweight, digital cameras that can record directly onto large hard disks, which no longer significantly limits the amount of footage that can be recorded at once, and digital editing systems, that allow now for the literally seamless blending of takes to create the impression of long, single shots. There had, of course, been films before the digital era that sought to create the impression that they consisted only of a long, single shot, and many more that have included long, virtuosic sequence shots, or extended sequences with hidden edits, often from a single character's point of view, like Robert Montgomery's *Lady in the Lake* (1947). The most famous, of course, is Hitchcock's *Rope* (1948). One would also have to include a film like Andy Warhol's *Sleep* (1963), which is a single long-take of a man sleeping, that lasts five hours and twenty minutes, or his even more epic *Empire* (1964), which is eight hours and five minutes in length (depicting six hours and thirty-six minutes of duration, though, as it is a slow-motion shot, filmed at 24 frames per second, but meant to be projected at 16). The filming of both, though, required reel changes—as Eisenstein tells us a *film* of any significant duration must, given the finite length of the filmstrip—and are, in that respect, not true single-shot films, requiring at least a minimal degree of "editing," when the long strips on the reel are eventually and necessarily spliced together, before they pass through the camera, as a result of the linear limitation.

While the earliest films, in the first decade of the cinema, were, for the most part, single-shot films, usually about a minute long at most, there have been later examples, when editing had become common, of *longer* single-shot films, such as Pascal Aubier's *Le Dormeur* (1974), which is just over nine minutes long—that is, about as long as a reel of film that could be loaded in a camera magazine. Béla Tarr's *Macbeth* (1982) is a full, feature-length single-shot film from the pre-digital era lasting about 72 minutes. But it was, in fact, shot on video as a TV movie for Hungarian state television, which is what allowed for the extended, unedited sequence. The film in fact consists of two shots, a five-minute pre-credit shot, and a fifty-two-minute shot after these and before the final credits. (There is a humorous fact on the IMDB entry for the film, where the "average shot length," calculated by adding these amounts and dividing by two, is noted as "1,860 seconds.")

These are quite exceptional examples, though—the quite rare instances of such single-minded (so to speak) dedication to the elaboration of very long shots, or entire single-shot films. The vast majority of films have been, and are still, even in the digital era, constructed with visible edits, even if that editing may otherwise be described as "invisible," to mean easily overlooked, according to the viewers' propensity to follow the inferential and causal logic, and the spatial and temporal coherence, that can create such a powerful effect of "continuity," or "invisible editing." Even if editing can in fact now be truly "invisible," the basic aesthetic phenomenon of montage, editing used explicitly, in relation to other techniques, to various formal, thematic, narrative, and depictive ends, remains intact.

5. The Art of Editing in the Digital Era

Abstract
In this chapter, I consider some of the specific stylistic developments in editing, as a result of the new affordances of digital technologies. The art of editing has developed in the digital era, but on the basis of earlier conceptions of the technique, in light of past stylistic developments, rather than as a direct effect of digital technologies, as the main determining force. The shift to digital technologies is presented as only the latest in a series of technical changes to editing, the stylistic effects of which are considered, as discussed by filmmakers and editors themselves.

Keywords: Editing, montage, film, style, technology

Editing had been seen as the last bastion of a traditional artisanal craft in an otherwise wholly technological art, the only moment in film production when the basic material was worked with directly, manually, by hand. With digitization, it might be argued, the cinema has lost its last enduring link to a pre-technological aesthetic, to the idea of art as a product of direct human control over a particular material or medium. The cinema has now been wholly incorporated into an advanced technological system, the system of computing, that has transformed so much of contemporary culture. Cinematic imagery and sound are converted into abstract, numerical, digital data, distancing the film artist from the material aspects of the medium. It is this sense of distance that has troubled some. This was a common complaint among many editors and filmmakers when new electronic and digital systems were first being introduced. No longer able to handle the film directly, there was concern that this would lead to a loss of control, to the complete abstraction of the process. The renowned editor Walter Murch, whose career spans the digital divide, has described what he calls the "unbearable lightness" of non-linear digital editing. In his short book, a meditation on the art of film editing, Murch writes that it began as a simple process for the direct manual manipulation of exposed film stock,

Furstenau, M., *The Aesthetics of Digital Montage: Film Editing and Technological Change.* Taylor & Francis Group, 2024
DOI 10.5117/9789463722803_CH05

as a bodily engagement with the physical material of the film. "In the first decades of the century," he says, "the film editor simply projected the uncut film, made notes, and then returned to a room equipped with a bench, a pair of scissors, a magnifying glass, and the knowledge that the distance from the tip of his nose to the fingers of his outstretched arm represented about three seconds."[1] Murch explains early editing in the homely terms of handicraft, describing the editing room as "a kind of tailor shop in which time was the fabric," where "the editor patiently and somewhat intuitively stitched the fabric of his film together."[2]

Originally presented as a lecture in 1988, as a summation of his experiences as a *film* editor, so much had changed by the time it was published in 1995 in book form that Murch was obliged to add a lengthy "Afterword," considering the effects of digital technology, what he characterizes as "the sophisticated assault of a number of electronic systems that have emerged over the last decade or so."[3] Many editors, he notes, resisted the change, in the same way that the introduction in the 1920s of the Moviola, the first dedicated mechanical system, explicitly designed for film editing, was resisted by those early editors who had been simply "stitching" their films together, in a literal sense, working by hand, with scissors and glue. "It is startling to recall," he says, "that the humble Moviola—that frog-green fixture of every editing room over the last seventy years—was initially rejected by those pioneering editors as expensive and cumbersome, even dangerous. Still worse," he says, from their point of view, "its main feature—the ability to study the *motion* of the images before deciding which frame to cut on—was dismissed as a crutch that would simply get in the way of the work to be done."[4] The concern, even then, was that with such an elaborate device—when compared to just a pair of scissors and a pot of glue—the task of film editing was being abstracted from the material reality of the filmic medium. Despite this development, though, and all subsequent technical advances that would follow, editing was still reducible to the physical fact of cutting strips of film. It was that basic fact that was to change with the introduction of the computer.

Murch's first encounter with computerized editing was in 1968. He was working with Francis Ford Coppola, who, with George Lucas, was looking

1 Walter Murch, *In the Blink of an Eye: A Perspective on Film Editing* (Los Angeles: Silman-James Press, 1995), 75.

2 Ibid.

3 Ibid., 76–77.

4 Ibid., 75–76.

into an early system called CMX—a prototype electronic "offline" videotape editing system that had been developed by a corporate joint venture between CBS and Memorex. "The potential of what I saw," says Murch, "naively made me think that this would overwhelm the industry in, at most, five years."[5] He and Coppola had intended to edit *The Godfather* (1972) on the CMX system, he says, "but nothing came of it: storage and access were still too limited for the amount of footage generated by a feature film."[6] But for *Apocalypse Now* (1979), Coppola tried another electronic, linear video-editing system, "on which he could experiment with different story structures."[7] Murch used the system "to compose the series of overlapping four-element dissolves in the first reel of that film."[8] The rest of the film, though, he notes, was edited on traditional film systems, including, surprisingly, a Moviola, which had, by then, come to be seen as an old-fashioned apparatus that had been more or less replaced by more sophisticated and flexible flatbed systems, such as the Steenbeck or the K-E-M, but which even Murch still liked to use. Murch would soon leave the Moviola behind, though, and even the more common flatbed editor—the whole traditional, physical apparatus of film editing—as he began to embrace the new computer systems. He worked again with Coppola, on *Godfather III* (1990), a good part of which was edited on the (aptly, interestingly named) "Montage" system, which Murch describes as one of the first truly "random-access, computer-controlled"[9] editing systems, marking the transition for Murch away from physically editing or "cutting" film and towards truly electronic, non-linear editing with the new digital systems.

By 1995, Murch was more or less convinced that the future of editing was electronic and digital, despite some lingering reservations. His tendency, he says, "as early as that first encounter with the CMX in 1968, was an enthusiastic but naive acceptance of electronics in the editing room—it seemed natural and inevitable to me then and, on a certain level, it does now."[10] Still, he says, he has "a respect for much of the inherent (and some of the unwitting) wisdom of mechanical editing, which has been slowly accumulated through many thousands of man-years of practical experience by the editing community."[11] What Murch, like many of his generation

5 Ibid., 78.
6 Ibid.
7 Ibid., 79.
8 Ibid.
9 Ibid.
10 Ibid., 112–13.
11 Ibid., 113.

of editors, thought would be lost, as he says, was "the plain fact of film's physical existence—that the image is actually captured, chemically, on celluloid."[12] This fact seemed significant to Murch, who tries to describe the relation between the materiality of film and the ephemeral nature of the cinematic experience.

> What we are communicating through film is so insubstantial—fleeting emotions patterned with beams of colored light and waves of sound—that it is somehow restoratively balancing to be able to actually hold a frame of film in one's hand. There is a wonderful alchemy in this: that hydrocarbons and silver and iron and various colored dyes, plus intangible time, can transmute themselves into the stuff of dreams.[13]

It seemed important to Murch, that is, given the fundamentally insubstantial nature of the cinematic experience, that there be a stage in its production when something material could be grasped, when the fact of manipulation could be felt directly, made substantial, when the medium could be confronted in its concrete tangibility, its materiality. He goes on to say that "compared to the stream of a computer's weightless electrons and ghostly digits, film is gravity itself."[14] While he admits that digital editing offers unparalleled speed and flexibility, he wonders if, before the process is comprehensively digitized, "we cannot use the lingering grace period to take another look at film's physicality and find a way to incorporate, somehow, various simulations of the best of that physicality into the new systems. A reel of film," he says, "*is* obstinately the way it is and no other—unless you actually rearrange the matter out of which it is made—and there is an anchoring, inertial value in that which the fluttering electrons, for all their nervous power, do not share."[15]

It is not entirely clear, though, that the contrast between film editing and digital editing is as significant as Murch suggests, or that the difference is reducible to that between hydrocarbons and electrons, that such a contrast is even a meaningful one. The "intangibility" of the cinematic experience is not obviously related to the specific physical compounds used in the production of the imagery—at least not in any obvious or clear-cut manner. In both cases, material systems are used to create images to be projected, or

12 Ibid.
13 Ibid.
14 Ibid.
15 Ibid., 113–14.

displayed on screens, as the basis for an imaginative, conceptual, interpretive experience on the part of the viewer. The ephemeral "intangibility," in this sense, is precisely what explains the cinema as an aesthetic experience, the cognitive and critical engagement of the film viewer, but in a way that is in fact characteristic of any art. The material object of the book, for example, is only the physical basis for the "intangible" experiences of the reader; a painting is the material basis for the ineffable imaginative engagement of the viewer; and so on. While the differences between the systems of production may be felt acutely by those creating the stories, scenes, or imagery, it is not clear that they are felt in the same way (if at all) by those experiencing the final product. Even for the filmmaker, moving from one technical system to another, it is not clear that the differences are as significant as often suggested. The bulkiness and material intransigence of analog film editing systems, one could argue, have merely been replaced by digital systems that are just as bulky, just as material, and it is not clear that one is more or less the source of creativity than the other. The materiality of digital systems, their physicality, is merely of a different kind. They offer new and undeniably more effective and efficient control over the moving imagery that is being shaped by editors, but through no less physical means. Rather than wielding scissors and glue, turning the handles of a Moviola, or scrolling the film through the reels on a Steenbeck flatbed editing table, editors now type commands on the keyboard, and click the mouse or trackpad, as they view footage through a software interface on the screens of their computers.

Murch was concerned that the material differences would make an aesthetic difference. Despite his original reservations, though, which he describes in his book, even having built his considerable reputation with the traditional tools of film editing, Murch was willing, and even enthusiastic, to make the shift to digital. With many more following his example, digital systems have become standard, almost wholly replacing traditional film editing, as Murch and most others have come to fully appreciate their benefits—or simply succumbed to their ubiquity. In any case, one rarely hears complaints now. Few filmmakers and editors are concerned about a putative loss of materiality. Few lament the loss of analog systems. In a popular handbook, *DV Filmmaking*, Ian David Aronson describes his first job as an editor—coincidentally in 1995, the year that Murch published his book—working with an analog, linear videotape editing system, an early electronic apparatus, that still had more in common with traditional film editing than the new, non-linear digital systems that were being developed. He describes the laborious process of finding and marking the points for cutting. While he could "preview" the edits (which he was in fact discouraged from doing, since at the time he was, as he says, "being paid by the hour"), once they were made, they were effectively irreversible, "because analog video editing didn't allow going back to make changes. ... Today," he says, "to the joy of editors around the world, adding or removing a shot requires nothing more than a few keystrokes or a mouse click."[16] Many of the benefits that Murch, despite his misgivings, already saw—increased speed, reduced costs, easier access to material, improved working environment, greater integration with other aspects of production, including the creation of sound and visual effects—have been consolidated, making digital editing the most common choice among filmmakers now. The many problems that Murch saw in 1995—insufficient storage and memory, poor display, sync problems, and many others, both technical and creative—have been addressed for the most part. As the benefits began to outweigh the problems, more and more filmmakers were willing to adopt new digital systems.

Still, one might argue that the transition to digital editing has meant that a once simple task, possibly the simplest of all the various tasks involved in filmmaking, has become as complex as any other—more complex, perhaps. Before the advent of non-linear digital systems, editing was one of the most technically straightforward aspects of filmmaking. The technical simplicity of editing has long been emphasized as its most distinguishing feature, but also as the source of its formative power, allowing the editor to focus on the creative tasks, rather than mastering complex technology and intricate technical systems. Karel Reisz, in his famous and influential handbook, emphasized the basic nature of the task. For a handbook, a practical manual, concerned with practical "techniques" for the working film editor, the book in fact includes very little technical detail. It is more of an extended essay on the aesthetic complexities of film editing, mostly concerned with style, offering a long account of the stylistic history of editing. Almost all of the strictly technical matters are in an appendix, a very brief, four-page summary that concludes the book, which emphasizes the technical simplicity of editing. This comes after more than three hundred pages considering various techniques of editing, the history of film style, in which editing is presented as a primarily aesthetic and formal undertaking.

Reisz's handbook has been re-printed many times, and was eventually updated and expanded, with a new section by Gavin Millar, added with Reisz's approval, to the second, enlarged edition, first published in 1968. This was to reflect the new editing and filmmaking styles that had emerged

16 Ian David Aronson, *DV Filmmaking: From Start to Finish* (Sebastopol, CA: O'Reilly Publishing, 2006), 89–90.

in the 1950s and 1960s, in light of some significant technological changes, including the development of lightweight cameras and the shift from the Moviola to flatbed editing systems. Millar describes the new filmmaking movements such as cinéma verité and the Nouvelle Vague, and the "personal cinemas" of filmmakers like Jean-Luc Godard and Michelangelo Antonioni, who were exploiting the new technologies to novel effect. The most significant technological change that is addressed in this later edition is not to the apparatus of editing, however, nor even the introduction of new lightweight cameras being used by young filmmakers, but rather the advent of widescreen processes, which were indirectly imposing new demands on the film editor. Reisz and Millar consider the formal and aesthetic consequences of the enlarged screen, from their perspective as editors. They point to André Bazin's warnings about "the dangers of manipulating reality by editing devices," as they put it, what they describe as "his mistrust of all that the unscrupulous—or simply thoughtless—artist can add to the image to distort reality."[17] Bazin, they argue, was to be credited with "popularizing" the idea, by then widespread, and encouraged with the advent of the widescreen, that editing was ideally to be kept to a minimum, that "the cut is in a sense an artificial device which draws attention in turn to the form rather than to the content," and that "the fewer cuts there are, the more 'natural' and spontaneous the action will appear to be."[18] They note that Bazin "makes no explicit reference to widescreen, then still in its earliest days," but argue that if his preference is indeed for dramatic composition within the frame, rather than by editing, by cutting, insofar as this is the "end" to which he thinks the cinema should strive, "it can be seen that the end that he is proposing is particularly well served by that process."[19]

17 Reisz, Karel, with Gavin Millar, *The Technique of Film Editing*, 2nd ed. (New York and London: Focal Press/Hastings House, 1968), 284.

18 Ibid., 283.

19 Ibid., 284. While Bazin indeed does not mention widescreen processes in the two essays that Reisz and Millar cite, "The Evolution of the Language of Cinema," and "The Virtues and Limitations of Montage," he does consider its effects briefly elsewhere, in a later essay, on the genre of the western, which would find considerable value in the size and shape of the new wide screen. Bazin, though, had little hope for the new process, or at least one patented version of it, considering what he describes as "the problem of CinemaScope." That specific process had been used in several notable westerns in the 1950s, which Bazin considers, including *The Kentuckian* (1955), directed by and starring Burt Lancaster, which, Bazin says, "bored the Venice festival to tears." The common view, he says, is that "while the enlarging of the screen is not called for elsewhere, the new format will renew the westerns whose wide-open spaces and hard riding call out for wide horizons. This deduction," though, he says, "is too pat and likely sounding to be true." The process, he notes, had in fact been better used in "psychological films such as *East of Eden*." He would, he concludes, "not go so far as to say that paradoxically the wide screen is

Widescreen processes, that is, seemed to promise the possibility of a more Bazinian cinema, with fewer cuts needed, as more detail, more elements of a scene, could be fit within the larger frame. As film editors, though, of course, Reisz and Millar are not willing to simply accede to Bazin's views on the matter entirely. They do acknowledge the opportunities provided by the wider screen, without entirely accepting Bazin's argument, but still granting it some credence. "However tenuous these speculations are," they note, about Bazin's account of the limitations of editing, "it is true to say that the wide screen has certain distinct advantages. Its width and clarity make it possible to unfold a scene of great complexity and length without losing sight of detail or overall shape, meanwhile preserving the intensity which is dissipated by cuts which are only mechanically necessary."[20]

Nevertheless, the new possibilities created by the wide screen should not, they argue, be seen as a kind of triumph over editing, which retains its value. Indeed, the new processes in fact offer as many new possibilities for editing as they do for *mise-en-scène* and the formal composition of the shot itself, emphasizing the continuing relevance and significance of editing in the widescreen era. Reisz and Millar are willing to acknowledge the new creative opportunities that the specific development of widescreen technologies could offer, but they also saw the new possibilities for creating some of the familiar effects of editing, but in the large-scale compositions of widescreen, as well as some new effects. They insisted, though, that editing was not necessarily a competing value compared to complex widescreen composition, which was certainly different but no better or worse as a technique than the creation of cinematic sequences with cutting. The new, larger, and more rectangular shape of the screen allowed for considerably more detail, and thus for new compositional arrangements with more visual elements. As Reisz and Millar say, the "wide screen is particularly suited

unsuitable for westerns or that it adds nothing to them, but it seems to me an already accepted fact that CinemaScope will add nothing decisive to the field." See André Bazin, "The Evolution of the Western," in *What Is Cinema? Vol. II*, trans. Hugh Gray (Berkeley: University of California Press, 1971), 157. Bazin's objections seem to be to CinemaScope as a particular process, which did indeed have some technical problems, as John Belton has noted in his study of widescreen. His objections may well have been more principled, though, as Belton in fact suggests, in an effective critique of Bazin's argument. The kind of history offered by Bazin, and others like him, whom Belton describes as "idealists," was, he says, unable to capture "the complex interplay of ideological, technological, economic, and sociocultural forces that informed the development of widescreen cinema." Moreover, Belton adds, the "appeal of widescreen in the 1950s rests as much on its production of greater spectacle as upon that of greater realism." See John Belton, *Widescreen Cinema* (Cambridge, MA: Harvard University Press, 1992), 2.

20 Ibid., 285.

to diagonal or horizontal compositions,"[21] as opposed to the more compact compositions of the more or less square frame of the traditional film image, on the basis of which standard editing patterns were established. Traditional editing, though, Reisz and Millar argue, was still possible in widescreen, despite the change in scale and shape. Citing a number of examples from films of the time, they argue that, for instance, the "size of the screen is not a hindrance to the big close-up."[22] Widescreen composition, they insist, can in fact give added intensity to traditional compositions and editing patterns. "A combination of the moving camera, changes of scale, and the exploitation of the screen's full width," they say, "can achieve effects of unusual power or delicacy."[23] The most significant observation they offer as editors, perhaps a kind of concession, is that widescreen composition may reduce the overall amount of editing necessary. They note that "the combination of widescreen and camera movement together ... produce an effect which makes unnecessary the cut one might have looked for in the past."[24] Yet editing remains viable, just as significant as a technique, even in the widescreen era, while the widened screen nevertheless offers the filmmaker the choice to reduce the number of cuts.

While such technological developments were introducing some incremental changes to editing style, the technology of editing itself was more or less the same simple, mechanical process it had been when Reisz had originally written the book, and that it had been since the invention of cinema itself. While advances in other technical aspects of the cinema could be accommodated by editors, even exploited for more powerful effect, Reisz and Millar still effectively implied that editing itself should ideally remain as technically simple as possible. The transitions that had taken place, as the Moviola was replaced by more complex flatbed systems, for example, had in fact changed little about the editor's basic tasks, little about the basic technical fact of editing. By the 1970s, though, a major transformation was already underway that was to profoundly alter the basic technology of editing. The first signs of change were to be seen in television production, where the electronic editing of videotape had become standard. Thomas Ohanian describes the many and very different electronic non-linear editing systems that had already become available by then, in his handbook for editors in the digital age, one of the first and most influential to be published. Editing

21 Ibid.
22 Ibid.
23 Ibid., 287.
24 Ibid., 296.

had become part of complex computing systems, making it very different, and far less simple, than what was undertaken with the technical apparatus for traditional film editing. "There is no computer in a film flatbed," he says, "just a series of motors, and film and magnetic track transports. When something goes wrong, it is usually a failed braking mechanism or a burned out bulb."[25] The new computer systems, by contrast, were much more difficult for an editor to understand. The inner workings were obscured from view, and required a specialist knowledge to understand, much less repair. Despite such technological complexity, though, Ohanian and others were for the most part enthusiastic about the benefits of the new systems. Most significantly, as Ohanian says, the "promise of digital nonlinear editing is that it will bring together not only film, video, and audio, but also a variety of other media that have never had one common environment in which to coexist."[26] He provides a list of benefits, in addition to this "ability to integrate different media easily," that include "creative flexibility," and "savings in money and time," and, more broadly, "preparation for digital integration," when the cinema as a whole, that is, will take digital form—a development that Ohanian was anticipating. The digital editing system would then be the hub, the fulcrum, so to speak, of digital filmmaking, where the various elements are made available, in a common form, for the filmmaker to work with.

The transformation, though, despite the kind of enthusiasm that Ohanian and others expressed, was in many respects a profound one, altering many of the most familiar aspects of the job of film editor. While so little had changed in the basic mechanical apparatus of film editing in the first ninety years of film history, in less than a decade it had been almost wholly transformed reconstituted in an entirely novel technical form. Such wholesale technical change would surely have significant formal and aesthetic consequences. Earlier technical developments, though, had not seemed to change anything fundamental about film art. While scissors and glue had long been replaced by more intricate and complex, and specifically designed, purpose-built, mechanical systems, the basic tasks of editing remained straightforward. From the Moviola to the Steenbeck, filmstrips were wound on wheels, displayed on small, rudimentary viewing screens, and physically cut with a blade and joined with glue or tape. By the early 1990s, though, this basic procedure was quickly being displaced by computing systems that dispensed

25 Thomas Ohanian, *Digital Nonlinear Editing: New Approaches to Editing Film and Video* (Boston and London: Focal Press, 1993), 298.
26 Ibid., 2.

with the material necessity of cutting and splicing. There were many very different non-linear digital systems becoming available that editors had to choose from, and while there were specific differences among them, they all manifested a profound change to the basic tasks of editing, which were no longer strictly mechanical. Once film imagery was available in the form of digital information, there was no end to the possibilities for designing systems for the display and manipulation of that information, and, for the first time in the history of editing, there was a quite wide variety of options.

It is this *plurality*, as we could call it, the range of choice of devices and systems, that was new. Editing technology to that point had been quite *singular*. Almost all editors, at any particular time in film history, used basically the same apparatus. While the Moviola was once the dominant device, it was eventually replaced by the various flatbed systems. Even then, only a few basic and similar models were available, and the Steenbeck was the most popular. In the digital era, though, there were many more competing devices. While the film editing systems were constrained by the physical fact of the linear filmstrip, which all systems had to run through the apparatus in basically the same way, digital image information could be rendered in any manner conceivable, displayed and made available for manipulation in very many different ways. Ohanian describes videotape-based systems, such as BHP TouchVision, Ediflex, and Montage (which Francis Ford Coppola and Walter Murch had used when editing *Godfather III*, in 1990); LaserDisc systems, like the CMX 6000, EditDroid, and Epix; and the purely digital systems, including the very popular and influential Avid system, but also DVision, EMC2, and Lightworks. He concludes his book with descriptions of then brand-new digital audio and video formats for personal computers, such as QuickTime, but also higher-quality DVI and MPEG formats. A transformation had occurred, so that, as Ohanian says, editing now meant accessing imagery in digital form, through "a computer display showing footage while the user interacts, not with machinery, but with computer software."[27] While an editor could choose the system that provided the most intuitive and simplest access to, or display of, the "footage" being edited, the software system behind the interface was vastly more complex than the earlier mechanical devices.

For the most part, though, digital editing systems, while technologically much more complex and various than mechanical systems, have in fact been designed to make editing seem as simple as it has always been, to allow for the editor and filmmaker to focus on the formal and aesthetic tasks at hand.

27 Ibid., 298.

Little seems to have changed in this respect. Still, there is a widespread sense, among filmmakers and editors, as well as many film theorists, that the shift to digital editing has changed something fundamental about the very nature of the cinema, and that the change has something to do with the altered material state of the medium. What that change is, though, is difficult to describe. Murch, of course, in his efforts to account for the differences between film and digital editing, had offered a false comparison. The chemical compounds of the celluloid emulsion, "the hydrocarbons and silver and iron," which he describes as reassuringly physical, and which he seems to privilege, are no more so than the "electrons and digits" that he thinks are so "ghostly" and insubstantial. And, in any case, he is describing only his experience as an editor. For the film viewer, the cinematic experience remains largely unchanged, whether created chemically or electronically. John Belton has made this point most forcefully, arguing that the effects of digital technology have had little effect on film viewing. In an essay originally published in 2002, Belton distinguishes between the effects of technological change on production and distribution, which may well be described as revolutionary, and the effects on the experience of film viewing, which are, in aesthetic and formal terms, negligible, if not outright inconsequential. He says that

> digital cinema is a revolutionary technological innovation for filmmakers like [George] Lucas and for the interests of corporate synergy that currently drive Hollywood. ... [I]t is also a potential boon—in the form of cost saving—for film distributors. But it is not yet clear that it can do anything for motion picture audiences aside from eliminating jitter, weave, dirt, and scratches from the projected image. Even if we concede that these improvements result in better projection, they are not significant enough for them to be declared "revolutionary" in terms of the audience's experience of motion pictures.[28]

Belton is mainly talking about projection. Can we in fact say the same thing about editing in the digital era? The changes to editing, and to the nature of the film image itself, making it more comprehensively amenable to a more expansive mode of editing, seem more significant, and many film and media theorists argue, like Murch, that a more basic and fundamental change has indeed occurred with the transition to digital technologies, that the very nature of the cinema has been altered, that with digital editing

28 John Belton, "Digital Cinema: A False Revolution," *October* 100 (Spring 2002): 107–8.

the last vestige of cinema's materiality has been lost. While Murch, as an editor, is reconciled to the change, given the benefits the new technology seems to offer, many film theorists, as we have seen, have argued that the change is profound, and that it has meant the loss of what had essentially defined the cinema as an art. The digital cinema, and digital editing, are often understood as distinct from, and as constituting a fundamental break with, the history that precedes them. The impression of a radical break, or a profound transformation with the introduction of digital technologies, is an effect, though, of a specific historical understanding of the cinema and of the effects of technical change. Histories are often informed by certain basic theoretical assumptions about the nature of the cinema, about its material or technical basis. If changes occur that seem to alter that basis, or to incorporate material elements that seem alien to it, there is a temptation to declare that what follows is wholly different from what came before.

***

Editing in the cinema *has* changed, though, in significant ways, having become almost comprehensively digital. It is rare, and will soon likely be impossible, for a film editor to physically "cut" film. Images and sound are now either recorded with a digital camera, or they are shot on film and then converted to digital information, and edited on a computer, "virtually," in the form of audio-visual information rather than material film-stock. The traditional tools of film editing—first scissors and glue, then the Moviola, the Steenbeck, and all the physical systems for cutting and reassembling pieces of film—have been replaced by software, by computer programs, which have transformed what was once an essentially "linear" process into a "non-linear" one. Films no longer consist of strips of film—*lengths* of film—a celluloid base supporting a photo-sensitive chemical emulsion, that have to be wound on spools, cut into shorter strips, and glued back together, in what was a laborious and time-consuming process—everything running, so to speak, in only two possible linear directions: forwards or backwards. The material being edited now, the digital images and any accompanying sounds, may be accessed "randomly" and instantly, called up at any moment and compared on the screen with other images, connected with other sounds, assembled, re-assembled, altered, manipulated—"processed" digitally, we might say, as rapidly as one may wish, without the need any longer to scroll back and forth, through long, linear, ribbons of film. The "film" material, once experienced primarily in terms of *length*, exists now in the non-linear *space* of the computer.

One may, moreover, do much more than just "edit." Other elements may be easily added—sounds, of course, and music, but also other pictorial components, either photographic or computer-generated, graphic material, titles, designs, and so on. Digital software allows for the consolidation of many of the various tasks of filmmaking, that had been undertaken separately. Editing may now be thought of as a complex and comprehensive process of synthesis, or, as I have been suggesting, as a system for the management and organization of immense amounts of visual and auditory *information*; not just the arrangement of shots in a sequence, but the organization of all the possible elements that will comprise the shots—to the extent, perhaps, that we may no longer be able to think of the shot as the most basic element in the creation, or in our understanding, of a film. Filmmakers, editors, may now work at what we could call a more "granular" level, with smaller, more fine-grained cinematic elements. While filmmaking had always included various processes for "compositing" discrete elements into a single image—optical printing, superimpositions, etc.—it was inevitably limited by the materiality, by the specifically physical fact, of the photographic image, which could be manipulated only, so to speak, on the surface. Editing was, in this respect, restricted, one could say, by the very limits of photography itself. Such restrictions seem to have been overcome, and editing is now perhaps best understood as an extensive set of procedures for the arrangement of images in sequences, but also for the assembly of the image in the first place—for the arrangement of elements within the shot and not only the arrangement of shots.

The question is whether this is a significant change or not. For film editors it has seemed a momentous, and often welcome, change, at least as far as the effects it has had on the way that they do their job. Ken Dancyger makes it clear that the changes have been profound. "Today we no longer speak of the coming technological revolution or how digital technology will transform sound, special effects, film, and video editing. The revolution is here. ... The implications," he says, "are enormous."[29] The revolution had long been anticipated, quite eagerly by some. In 1991, the film editor Craig McKay, who'd worked on such films as *Reds* (1981), *Something Wild* (1986), and *Silence of the Lambs* (1991), was asked in an interview by Vincent LoBrutto: "In the future, what medium do you think film editors will be working on?" McKay answers: "I think down the line that digitized images are certainly the future of this business. Maybe six years from now videotape will be out

29 Ken Dancyger, *The Technique of Film and Video Editing: History, Theory, Practice*, 5th ed. (Burlington, MA: Elsevier/Focal Press, 2011), 391.

the window; everything will be stored on laser disc, digitized and random accessed. I'm glad it's moving in that direction. That's definitely where it has to go."[30] Others seemed to agree. In the early 1990s, handbooks, technical manuals, guides and instructional booklets on non-linear digital editing began to be published in large quantity, as the traditional technologies and procedures were being transformed. The benefits were typically thought to be unambiguous and were spelled out quite clearly. As Dancyger says, "there is no question that nonlinear editing and digital technology will have positive effects on the editing process and on the outcome of that process, the screen story."[31] Thomas Ohanian lists the very clear and specific benefits he sees offered by the new technology noting that these are the result, though, of a more profound transformation, which he describes. "The promise of digital nonlinear editing is that it will bring together not only film, video, and audio, but also a variety of other media that had never had one common environment in which to coexist."[32]

The implications of Ohanian's observation are worth considering. Film, as a specific material, is understood here as only one medium among many in the cinema, which in turn is rather suggestively described as an "environment," the realm or space of artistic undertaking in which various media may be brought together for particular purposes and in order to realize specific artistic ambitions. In fact, digital technology is seen to have solved an implicit and longstanding problem, which is that there had actually been many different cinematic "environments" in addition to the specific context of shooting itself—pre-production, post-production, the various stages of which have been quite distinct if not wholly isolated from each other—within which particular media had to be dealt with on their own terms, eventually to be brought together only in lengthy, complex, and time-consuming processes, combined ultimately in a single final work. The promise of consolidation, of the creation of a single cinematic environment, is what most filmmakers and editors have seen as the signal achievement of digital technologies. Dancyger describes the various elements of production in the digital environment. "In preproduction, computer software is available for previsualization of scenes. ... During production, nonlinear editing allows for rapid assemblies that provide feedback. ... Digital cameras will replace film and electronic-based videotape as the originating source of the

30 Vincent LoBrutto, *Selected Takes: Film Editors on Editing* (New York: Praeger, 1991), 205.

31 Dancyger, *Film and Video Editing*, 392.

32 Thomas Ohanian, *Digital Nonlinear Editing: New Approaches to Editing Film and Video* (Boston and London: Focal Press, 1993), 2.

image. … In postproduction, it is possible for the editor to consolidate in his role, sound editing, picture editing, sound mixing, special effects, and printing … The editor could also write and input the music track using her nonlinear editing system!"[33]

Dancyger's enthusiasm is slightly tempered, however, wary as he is of the limits of any individual to manage all the various media, with the requisite skill. The creation of a single cinematic environment does not, he suggests, mean that filmmaking will or should cease being a collective, collaborative undertaking. "The degree of consolidation that is possible," he says, "is probably not wise for one person to undertake, but the critical point is that the digital revolution makes it possible."[34] The fact, though, that all media may now be brought within a single environment does not mean that an individual will necessarily have the talent to use them all effectively. Filmmaking will likely, and perhaps necessarily, remain a collective endeavor. The nature of the endeavor may well have to be reconceived, though. Given the common environment within which film production is now undertaken, it is easier for the various individuals involved to convene at virtually any stage in the process, and to contribute at various junctures when, in the past, they may not have been included in discussions. The traditional divisions of labor may well be breaking down, replaced by new models of collaborative filmmaking. It will likely also be the case—if it is not already so—that expectations will increase, that results will be expected sooner, that the entire pace of production will be sped up, with the assumption that it can all be done much more inexpensively. Many editors already bemoan the loss of what had been a somewhat insulated working environment—the editing room—where it was understood that things could happen only at a certain pace. Still, it is more common to hear about the benefits of easier and more immediate collaboration during the editing process, now that digital versions of draft cuts can be easily circulated and commented on, while others are working on the other aspects of the production. What remains, though, as Dancyger notes, is the need to make creative decisions about what to do with the various media that digital technology make available within a common environment, insisting that "a computer-driven editing machine such as an Avid or Lightworks station, no matter how sophisticated, cannot make the creative decision of where to cut and why."[35]

33 Dancyger, *Film and Video Editing*, 391.
34 Ibid.
35 Ibid., 392.

Again, such an observation is worth considering. While digital technologies will, Dancyger says, certainly make the process of editing, and of filmmaking generally, more efficient and less costly, and while it may allow editors and filmmakers to spend more time together making creative decisions—rather than, say, waiting for the processed film to return from the laboratory, or for the film to be spooled to the scene that is to be edited, or to be connected to another already assembled—the creative decisions themselves are no easier, nor are the new creative possibilities that digital technologies might engender any more readily apparent. One common assumption, says Dancyger, is that "the new technology will yield new forms of story-telling, new levels of interactivity, and a more democratic relationship between storyteller (film- or video-maker) and audience."[36] Yet he sees no evidence yet of any significant change along these lines, certainly not in the mainstream cinema. Interactive technologies, as he notes, have had a far greater effect on other forms, such as video games, and in the digital distribution of already existing works of art. In the cinema, it has become easier to incorporate ever more special effects into films, yet even the most spectacular digital effects films, such as, in the examples he provides, *Terminator 2: Judgment Day* (1991) and *Jurassic Park* (1993), remain "conventional screen stories, neither challenging old forms nor old ideas."[37] Those forms and ideas are manifested primarily in traditional editing patterns. Dancyger imagines that digital technologies may well eventually create new aesthetic opportunities. It may, he suggests, be easier now to imagine adapting more complex, psychological narratives, representing the interior lives of characters, which may be rendered more easily with digital effects, rather than merely depicting their actions in external spaces, which can be simply photographed. (Several films in the early era of digital editing seem to bear this out, from *Eternal Sunshine of the Spotless Mind* [2004] to *The Diving Bell and the Butterfly* [2007].) Far greater, though, than the revolutionary force of new technology, he argues, is the conservative force of familiar formal convention. While digital editing is now *non*-linear, he suggests, storytelling in most of the mainstream cinema around the world is still, importantly, linear. "And because the linear story, the plot-driven story, today reigns supreme, the likelihood is that digital technology will be used in support of linear tales rather than to subvert or emend narrative conventions of today."[38]

36 Ibid.
37 Ibid.
38 Ibid., 393.

Moreover, non-linear storytelling has always been possible with linear editing technology, apparent in the examples he gives, from Luis Buñuel's *Un Chien Andalou* (1929) to Orson Welles' *Citizen Kane* (1940). There is no obvious or necessary correlation, he insists, between specific editing and filmmaking technology and the kind of stories that can be told on film. Perhaps, he suggests, given that both linear and non-linear narrative have long existed, despite the former always being more dominant, the sheer fact of non-linear modes of editing may mean that filmmakers and editors begin taking more creative risks, using the new technology to experiment with narrative forms. This is not guaranteed, though, strictly by technical development. He distinguishes, helpfully, between non-linearity as "a technological fact," manifested in new digital editing systems, and as "a philosophical and aesthetic attitude,"[39] which may or may not be alive as a possibility in the minds of filmmakers and audiences. While the technical fact of non-linearity may have been achieved, it by no means ensures the acceptance of the non-linear narrative. Still, given the ease with which material may now be accessed, assembled, and reassembled on digital editing systems, he urges filmmakers to take advantage and explore the possibilities of unconventional, non-linear narrative. The possibility depends less on the editing technology, though, than on interested audiences, and on filmmakers who may be given the opportunities and the freedom to experiment, as working patterns start to change, as the complex contexts within which filmmakers work are restructured. In this respect, digital technology is far more significant insofar as it is changing systems of distribution and exhibition, lowering the costs of production, increasing the number of venues where films can be seen, making it possible to create many smaller audiences with more specific tastes and interests. "Linearity served its purpose," Dancyger says. "It will not disappear, but nonlinearity will now assert itself more aggressively. The digital age demands more narrative styles for the new but fragmented audience of the age."[40]

We shall see if Dancyger's prediction is borne out. More significant is the tempered quality of his and other accounts, by filmmakers and editors, of the effects of digital editing technology, and their usually quite careful distinctions between technological change and formal and aesthetic effects. Most are willing to imagine what new possibilities may be revealed as a consequence of technological developments, but they insist that whatever possibilities emerge, they will be the result of the creative efforts of

39 Ibid.
40 Ibid., 394.

filmmakers themselves, who will not be made more or less creative by the force of a particular technology. On the whole, most seem to welcome the changes, and are eager to see what might be done with it, without assuming that the technological change itself will have any direct or immediate effect on film style, or on the nature of the cinema as a creative endeavor.

Film theorists, by contrast, tend to be far less sanguine. "Technological innovations are often taken as harbingers of artistic doom,"[41] observes Noël Carroll, in a summary of the history of responses by film theorists to technical developments. It has been common, throughout the history of the cinema, to see whatever new change has occurred as antithetical to the very nature of the art, or to assume that a new technique—the advent of synchronized sound, for example, or color photography—as an obstruction to any further aesthetic development. In such cases, a particular aspect of the medium—or, as I have suggested, a particular medium from all the various media that can be used within the environment of cinematic activity—is nominated as the most fundamental, and any change is seen as a potential threat to that aspect, and therefore as a threat to the cinema itself. What Carroll describes as the regular "dire forebodings" of film theorists and critics, though, "have consistently come a cropper. The problem with such aesthetic jeremiads is that the theorists who propound them fail to attempt to imagine the ways in which new developments in technology may bring in their train new possibilities. We need to be thankful that we have artists to do that."[42] It is also important, as I have been arguing, to recognize that new developments in technology also tend to consolidate already existing formal and aesthetic achievements, for which we also have to be thankful to artists who navigate the often uncertain terrain of constant technological innovation to provide familiar formal experiences to audiences.

In any case, it is worth contrasting the discourses of production and of reception. Filmmakers and editors tend to consider what one may do with a new medium to *determine* its aesthetic value, while theorists typically look to the medium itself to *discover* that value. Lucy Fischer argues that digital technologies allow filmmakers to do something new, to edit not only *between* shots, but now also *within* them.[43] She concludes, as I have noted, that this will potentially change the "ontology" of the cinema as an

41 Noël Carroll, "Film as Art: Introduction," in *Philosophy of Film and Motion Pictures: An Anthology*, eds. Noël Carroll and Jinhee Choi (Malden, MA and Oxford: Blackwell, 2006), 16.
42 Ibid., 17.
43 Lucy Fischer, "Film Editing," in *A Companion to Film Theory*, eds. Toby Miller and Robert Stam (Malden, MA and Oxford: Blackwell, 1999), 81.

art. But the point is that, as Dancyger in fact implies, it is more a question whether filmmakers will choose, now that they can, to edit "within" the shot, or continue, as many still do, to edit "between" shots—or a combination of both. In either case, cinematic art remains the same, understood as an inevitably changing technical and aesthetic environment within which film artists make all kinds of choices. Fischer is confusing what is a new choice available to film editors and filmmakers, to be placed within the context of the many possible choices that are available to them, with a supposedly normative property of a medium. For Fischer, the new ability to edit "within" necessarily comes at the cost of some presumed integrity of the photographic image. She assumes that the original difficulty of getting "into" a photograph to alter it implies an inherent value for such integrity. In the face of significant technological change, and of the advent especially of digital editing, many contemporary theorists have declared that the traditional cinema, which is thought to have possessed a uniquely "photographic" value, has been comprehensively transformed, and that a new kind of theory is needed to account for what is assumed to be a wholly new phenomenon, the digital cinema. In this respect, they are echoing certain tendencies in earlier film theory, as I have argued, when what we might call the first technological change occurred, that is, when the cinema itself was invented, and the theoretical task was understood to be the justification of the medium as of potential aesthetic significance, but also as distinguished from the other traditional arts. The first question asked by most film theorists was whether film was, or could be, art. The answer tended to be, either, that it was, and that it could be understood in terms of the features and qualities that it possessed that it shared with the traditional art, or, more commonly, that it was a new kind of art, which would disturb if not destroy the traditional concept of art.

Contemporary accounts of digital technologies in the cinema, and specifically of digital editing, can be compared, with some careful qualifications, of course, to the familiar concerns of earlier film theorists in the face of other technical changes, most significantly the coming of sound, and the introduction of color photography. It can be compared instructively, as well, to what had often been proposed as the first fundamental task of film theory, which was, namely, to justify film as art, to provide the theoretical justification thought necessary for film to be taken seriously as an art, to explain the nature of the new medium as it was first emerging. Hugo Münsterberg, Béla Balázs, and Rudolf Arnheim are among the most significant of the early film theorists who saw the main task of film theory as a justification of film as art, of demonstrating, by philosophical argument, that film was capable of serving traditionally aesthetic ends. Münsterberg asks, in *The Photoplay*, in a rather tentative tone, "do the moving pictures bring us an independent art, controlled by independent esthetic laws of its own, working with mental appeals that are fundamentally different from those of the theater, with a sphere of its own and with ideal aims of its own?"[44] He answers, by way of a neo-Kantian analysis of the relations between representation in the "photoplay" and the basic structures of human perception and mental activity, that it *does*—although it is a statement of cautious hopefulness, given what he conceded were the rather meagre aesthetic accomplishment of the "photoplay" to that point. Arnheim is more confident, simply stating, in the famous opening lines of *Film as Art*, that film, "like painting, music, literature and dance," he says, "is a means that one may, if one wishes, use to create art."[45] He then provides a detailed justification of film as an art, which, he says, and as Münsterberg had also insisted, does not merely "reproduce reality mechanically," but creates something new. Unlike Münsterberg, though, Arnheim argues that film is radically different from perception, arguing that "it is utterly false for certain theorists, and some practitioners, of the motion picture to assert that the circumscribed picture on the screen is an image of our circumscribed view in real life. That is poor psychology."[46] As he insists, the supposed "limitations" of the cinematic image cannot be understood as similar to and modeled on the "limitations" of visual perception, and cognition, "because in the actual range of human vision the limitation simply does not exist. The field of vision is in practice unlimited and infinite."[47] Films are, as he says, quite simply, "pictures," not recreations of natural vision. And it is the case that "the limitations of the picture are felt immediately," concluding that "it is just such restrictions which give film its right to be called an art."[48]

44 Hugo Münsterberg, *The Photoplay: A Psychological Study* (New York and London: D. Appleton and Co., 1916), 43.

45 Rudolf Arnheim, *Film als Kunst* (Suhrkamp, 2002 [1932]), 24; my translation. In German, Arnheim uses the word "*Mittel*," to describe film, and the other arts he lists, which can be translated either as "medium," related to the German adjective, *mittel*, or middle, or in a more general sense as "means," as in the term that Arnheim uses elsewhere, *Kunstmittel*, or "artistic means. "Mit dem Film steht es ebenso wie mit Malerei, Musik, Literatur, Tanz: man kann die Mittel, die er bietet, benutzen, um Kunst zu machen, man braucht aber nicht." The published English translation is: "Film resembles painting, music, literature, and the dance in this respect—it is a medium that may, but need not, be used to produce artistic results." See Arnheim, *Film as Art* (Berkeley: University of California Press, 1957), 8.

46 Arnheim, *Film as Art*, 16.

47 Ibid., 16–17.

48 Ibid., 17.

In both cases, though, the two very different, indeed opposite, claims are in the service of the same goal, to establish film's aesthetic *bona fides*. And both make the same conceptual error, which is to assume that a medium may be used artistically only once its inherent aesthetic properties or propensities are discovered and described, with the consequent corollary—which is sometimes stated explicitly, other times left implicit—that such artistic use is to be guided by such theoretical knowledge, that the medium's inherent properties are to determine the artist's efforts. While Münsterberg assumes that such a propensity is to be found in the medium's ability to reproduce the mental experience of perceiving the world, and for Arnheim that it is in the difference between perception and the film picture, in both cases they have supposedly found not only what they think is the inherent aesthetic criterion of the new art form, but they both present it in normative terms. The film world, as Münsterberg puts it, is ideally meant to be nearer to the "mental world." Its pictures and techniques should properly build upon and acknowledge the relations between cinema and thought. For Arnheim—and this explains his adamant refusal, even decades after the fact, to accept that synchronized sound and color could be aesthetic components of the cinema—films should strive to maintain, and indeed emphasize, their *difference* from reality. Sound and color will merely cause cinema to revert to its non-aesthetic status as a recording mechanism—providing more and more detailed copies of reality, of our perception of reality, in all its chromatic and aural diversity, verging on incoherence and confusion when rendered artistically as a too detailed picture.

Other examples of strictly prescriptive theories of film, based on an analysis of what is *basic* or essential to the medium, can be easily found. Siegfried Kracauer, for instance, argued influentially that film is basically photographic, and that the cinema can be separated into those films that acknowledge that fact, what he calls the realistic tendency, and those that try to resist or overcome it, what he calls the formative tendency. His account of the "basic concepts" of film, which is derived from an initial analysis of the properties of photography, concludes with the issuing of a normative edict: "It follows from what has been said in the preceding chapter [on photography] that films may claim aesthetic validity if they build from their basic properties; like photographs, that is, they must record and reveal physical reality."[49] André Bazin, of course, is perhaps most famous for declaring, in his essay "The Ontology of the Photographic Image," that the

49 Siegfried Kracauer, *Theory of Film: The Redemption of Physical Reality* (Princeton: Princeton University Press, 1960), 37.

cinema is based in and derives its artistic identity from photography, which is the "automatic means" now available to not merely represent an object, but to "transfer" something of it into the image itself. "Photography," he says in his later essay "The Evolution of the Language of Cinema," "enjoys a certain advantage in virtue of this transfer of reality from the thing to its reproduction."[50] Like Kracauer will do later, he makes a distinction between those who build upon this advantage, and those who ignore it, in his famous formula according to which there are "those directors who put their faith in the image and those who put their faith in reality."[51] It is the latter, of course, who, according to Bazin, have the more legitimate claim to the status of film artist, given their fidelity to the supposedly basic nature of the medium.

Münsterberg's and Arnheim's theories are more or less out of favor today, partly, it would seem, because they are so obviously prescriptive. (Although one might argue that a great deal of cognitive film theory explores the supposed relations, or dissimilarities, between cinematic representation and human perception and cognition, developing the formalist, even Kantian, quality of both theorists.) Yet other similarly prescriptive accounts of cinema, including Kracauer's, and perhaps especially André Bazin's, have remained popular—in Bazin's case, as George Wilson had noted, likely because there is much that is otherwise so incisive in his arguments, if they can be separated from his more prescriptive assertions.[52] Although it is also important, as Wilson says, that Bazin may well have intended his analysis as properly prescriptive, thus limiting the full power of his account of cinema. The continuing influence of both Kracauer and Bazin is the result, I think, of the powerful analyses that they provide of the effects of editing, or montage, the power of which both acknowledge, but which both think should be attenuated in favor of the more genuinely cinematic features of filmic representation which are presumed to derive from their

50 André Bazin, *What Is Cinema?*, trans. Hugh Gray (Berkeley: University of California Press, 1967), 14.

51 Ibid., 24.

52 Wilson cites the "Ontology" essay. "On the whole," Wilson says, "Bazin seems to defend a metaphysical thesis of the 'intrinsic realism' of photography; that is, the idea that a photograph somehow 'shares its being' with the items of which it is a photograph. Though sometime he writes as though he were describing the 'impression' that he thinks we have in seeing objects in (many) photographs." It is this latter more limited view that Wilson could support, against the often more obscure aspects of Bazin's argument. "Whatever Bazin meant," Wilson says, "it is, of course, the 'psychological' version of his views that I find more plausible." See George Wilson, *Narration in Light: Studies in Cinematic Point of View* (Baltimore and London: The Johns Hopkins University Press, 1986), 212. n. 16.

photographic origin. For both, though, and in this respect, the question of editing is central, the use of which should be carefully judged in order not to interfere with the full and proper manifestation of the cinema as a *photographic* art. The effects of their arguments have, either implicitly or explicitly, shaped much of the discourse on digital editing.

Kracauer distinguishes between what he calls the "basic properties" of the cinema, which are, he says, "identical to the properties of photography," and the "technical properties." "Of all the technical properties of film," he says, "the most general and indispensable is editing. It serves to establish a meaningful continuity of shots and is therefore unthinkable in photography."[53] In addition to editing, there are the other technical properties, such as the close-up, double exposure, slow-motion, dissolves, all the 'special effects,' and so on, about which he will say little. "These scanty hints suffice," he says. "It is not necessary to elaborate on technical matters which have been dealt with in most previous theoretical writings on film."[54] Instead, he will concern himself with "cinematic techniques only to the extent to which they bear on the nature of film, as defined by its basic properties and their various implications."[55] So: "The interest lies not with editing in itself, regardless of the purposes it serves, but with editing as a means of implementing—or defying, which amounts to the same, such potentialities of the medium as are in accordance with its substantive characteristics."[56]

Bazin makes a very similar argument. His account of the photographic "ontology" of the cinema is perhaps more influential now than it has ever been, especially in accounts of digital editing, and of the digital cinema more generally. Bazin's well-known objections to editing, to montage, are the basis of several quite influential accounts of the consequences of digital editing, which are, on the whole, thought to be quite dire for the cinema, the triumph of a technique that Bazin thought should be tempered by an acknowledgement of the essentially photographic nature of the cinema. The photograph, Bazin says, in his well-known argument, allows for a significant interpretive freedom on the part of the spectator, given its causal origins, given the limitations that the medium places on the interventions of the artist. Editing, or montage, while inevitable, Bazin admits, should properly be subordinated as a technique to the primarily photographic identity of the medium. Editing, he argues, reduces the "ambiguity" of the image,

53 Kracauer, *Theory of Film*, 29.
54 Ibid.
55 Ibid.
56 Ibid.

which is what the film should naturally tend towards, as an effect of the photographic process. With the various techniques of filmmaking, but especially with "the resources of montage," Bazin says, "the cinema has at its disposal a whole arsenal of means whereby to impose its interpretation of an event on the spectator."[57] Filmmakers are judged by Bazin according to the degree to which they fully avail themselves of this "arsenal," which he clearly suggests is to be avoided, or whether they allow the inherent ambiguity of the medium to express itself. While he admits that editing may well be employed without destroying the importantly ambiguous quality of the imagery in a film, by maintaining a sense of the spatial integrity of the scene of action, the true flowering of cinematic art requires another technical development, namely the shot in depth. "It was this fashion of editing," which respected spatial integrity and the supposedly consequent ambiguity to the degree possible, says Bazin, "so admirably suitable for the best films made between 1930 and 1930, that was challenged by the shot in depth introduced by Orson Welles and William Wyler."[58] Depth of focus, he says, "reintroduced ambiguity into the structure of the image if not of necessity—Wyler's films are never ambiguous—at least as a possibility,"[59] a possibility that is, he argues, first fully realized in *Citizen Kane*.

Both Kracauer's and Bazin's argument are as susceptible to the criticism that, as I suggest above, could be made against Münsterberg and Arnheim, yet they are, especially Bazin's, being offered again as the basis of accounts of the advent of digital cinema, and specifically of digital editing. Traditionally, Bazin's and Kracauer's arguments have been understood to be importantly different from Münsterberg's and Arnheim's. While Kracauer and Bazin both emphasized and celebrated the work of filmmakers who focused on the "basic properties" of the medium, or who put their "faith" in reality, theorists like Münsterberg and Arnheim were seen as advocates for films that exploited the medium's "technical" or merely formal aspects. The distinction was given considerable force by Dudley Andrew, who, in his influential survey from 1976 of the history of film theory, divided that history into the "formative tradition," which included Münsterberg and Arnheim, but also Eisenstein and Balázs, and "realist film theory," manifested most acutely in the work of Kracauer and Bazin.[60] Yet the differences, as I have

57 This is from his essay "The Evolution of the Language of Cinema," in Bazin, *What Is Cinema?*, 26.

58 Ibid., 33.

59 Ibid., 36.

60 Dudley Andrew, *The Major Film Theories: An Introduction* (Oxford: Oxford University Press, 1976).

been suggesting, are more apparent than real, and almost all the early film theorists share a more fundamental goal, with "theory" understood primarily as a task of justification, as the search for the basic aesthetic qualities of the medium of film, which are then offered as the normative terms or prescriptive edicts supposed to guide any filmmaking endeavor.

Such arguments, as Noël Carroll has argued, were characteristic of early critical and theoretical writing on film, which began to appear almost as soon as the cinema had been invented, producing in very short order an astonishingly large literature. By 1988, when Carroll was writing, he guessed that film theory "comprises more titles than the theoretical wings of such ancient arts as dance and theater. … One reason for this," he suggests, "is that film was an art form that was invented not just in living memory," at least for those first, "classical" film theorists, "but self-consciously."[61] As a new medium, Carroll says, it had to "prove that it was an art form. … And the way that it initially set about getting itself taken seriously was to prove that it was an art—an art on par with its seven predecessors. This was the first task of film theory."[62] Yet this is a peculiar way of describing the series of events—the "self-conscious" invention of a new form, and the subsequent, and supposedly necessary, theoretical task of getting itself taken seriously, of being accepted as an art—which suggests a relationship between filmmaking and film theory that is not, I would argue, the right one. It may be more correct to say that some artists had, for whatever reason, and perhaps, one might say, courageously and hopefully, taken up a new medium, and endeavored to create works of art. Film "theorists"—writers, philosophers, critics, other artists—recognized their efforts as such, and then undertook to explain how such accomplishments had been achieved, or what the consequences of such achievements might be, or what might be learned through the contemplation of works of film art. They also, of course, felt obliged to defend the new art works, and the medium chosen to create them, to defend them from philistines who couldn't understand how art, in any traditional sense, could be created from such a new, "technological," medium. To do so, they typically sought to discover some inherent or essential feature of the new medium, which could be presented as part of their normative, justificatory argument. The "philistines" making such supposed objections are rarely if ever mentioned by name, though, in the early writings of film theorists. They are usually anonymous foils, against

61 Noël Carroll, *Philosophical Problems of Classical Film Theory* (Princeton: Princeton University Press, 1988), 4.
62 Ibid.

whose doubts about the aesthetic value of film arguments could be made for the specific, inherent values of the new form. Münsterberg simply poses the rhetorical question: "Do the photoplays furnish us only a photographic reproduction of a stage performance ... an inexpensive substitute for the real thing?"[63] It is supposed to be the sort of question that was abroad in the world of the skeptics. Arnheim merely insists that there are "still many educated people who stoutly deny the possibility that film might be art,"[64] (8), without saying who these people are, and what the nature of their objections might be. In both cases, the presumed presence of those who supposedly reject the claims of films to be art serves merely to provide the excuse for a search for the supposedly inherent qualities of the medium that will allow it to be used artistically, defying the naysayers, providing what was felt to be a needed theoretical justification of the medium as an art.

Film theory is still largely if more implicitly conducted as an act of justification, and still often proceeds as though it were a normative search for the essential quality of the medium. In recent years, the task has assumed a more melancholy tone, as theorists endeavor to show that what had once been the essentially photographic nature of cinema has been displaced, or replaced, by a new, digital identity. I would argue, though, that it is not up to film theorists to determine which of the many possible media that may be used to make a film, a work of cinematic art, is the most essential. Rather than a normative, prescriptive undertaking, film theory should be understood as a descriptive accounting of the cinema as a complex technological assemblage, with many different components, using a variety of media, none of which necessarily takes precedence over any other. The cinema is a technology—in fact, it is a collection of many disparate technologies, that can be used in various combinations to a wide and unpredictable variety of ends, and to perform many different functions. Among the many technological components that has comprised the cinematic apparatus, the tools of editing have been significant, and have undeniably undergone considerable technological change. The most recent change, with the advent of digital editing systems, though, is consistent with the technological history of the cinema, and is continuous with its overall development as a medium or technology of expression. To say so, though, is much easier than to explain how a particular art form, like the cinema, is "expressive" exactly.

***

63 Münsterberg, *The Photoplay*, 43.
64 Ibid., 8.

In a certain respect, the debates of film theory, in both the "classical" and "contemporary" eras, are about the limits on *expression*. Understood, by the realists, as an essentially photographic medium, the film artist was supposed to subordinate their expressive ambitions to the inherently, naturally expressive properties of the medium itself. The main way to do so was to limit the use of editing and to allow the image to fully manifest its own expressive significance. For the formalist, expression was what the film artists had to struggle to achieve, given film's inherently "reproductive" capacities, against which the film artist had to impose their own artistic powers. In both cases, though, the possibility of artistic expression is regulated by an assumed property of what is thought to be the primary medium of the cinema, namely photography. In both cases, the main expressive technique is editing, which should either be restrained or unleashed. Many have argued, usually from one or the other of these perspectives, that editing is "expressive," which I take to mean that the basic propensity for expression, which may be manifested in any medium, is manifested in particular ways in the cinema, ways that can be discerned by attending to the patterns that films take once they have been edited. If this sounds perhaps too enigmatic, or circular, it is an effect of the difficulty of saying how, precisely editing *is* expressive, or what expression itself is that it may be manifested in the form of film editing.[65] This complexity has been confronted most recently by Valerie Orpen, who argues that, in most accounts of editing, it is "taken for granted" that it is an expressive technique. "We all know that it is expressive," she says, "but it is more difficult ... to explain why and how."[66] Given this difficulty, Orpen argues, expressiveness is for the most part ignored in accounts of editing, which emphasize instead what she calls its "connective" function—the arrangement of elements into a coherent, connected whole, which can be understood in basic temporal, spatial and narrative terms. Seeking to identify and to describe the specifically *expressive* aspects of editing, Orpen analyses several notable films—*Rear Window* (1954), *Breathless* (1960), *Raging Bull* (1980). To be "expressive," she suggests, is to draw our attention to the fact of arrangement, rather than to merely create an effect of coherency, specifically spatio-temporal coherency, through arrangement. In each of her analyses she describes how the editing functions to draw the viewer's attention to the formal fact of the editing

65 This is the general problem of any theory of art as "expression." A good summary of the debate is offered by Aaron Ridley, "Expression in Art," in *The Oxford Handbook of Aesthetics*, ed. Jerrold Levinson (Oxford: Oxford University Press, 2003), 211–27.

66 Valerie Orpen, *Film Editing: The Art of the Expressive* (London: Wallflower Press, 2003), 3.

itself. "The editing in *Raging Bull*," for example, she says, "rests primarily on constant and noticeable change, change *intended* to be noticed, and this noticeability becomes a spectatorial pleasure in itself."[67]

The approach yields only modest results, though, given the challenge of identifying any particular aspect as representative of some distinctively "expressive" use of editing. Still, Orpen's is indeed one of the few attempts to isolate and examine editing as a specific formative element of the cinema, but while she calls editing "the art of the expressive," she does not offer an explicit definition of expression, a notoriously ambiguous term. Implicit in her argument, though, is an equation between visibility and expression, which is contrasted with invisibility and connection. When editing is functioning connectively, she argues, it is "invisible." We do not see the editing, but rather and only the connections that are made (or, rather, only the "whole" that is made by the explicit *connecting* of "parts"). When editing is expressive, though, it is "visible," that is, it is seen *as* editing, and *as* editing it is presumed to be expressive. Yet this suggests a too easy distinction between visibility and invisibility—terms that often have little more than a kind of figurative value in accounts of editing—as discrete conceptual categories that could be the basis of an analysis of the expressive value of editing. The effects of editing are manifestly "visible" in the overall form that a film has, and which the viewer can sense or appreciate, even if they do not see, or, more precisely, if they overlook the specific techniques, the actual cuts and other connective devices, that function ultimately to create the overall formal and thematic coherence. Indeed, editing is always "visible" (except, I suppose, in the rare cases of disguised cuts, as in Hitchcock's *Rope*, but even then ...). By "invisible" it is usually meant that the editing is so familiar, or that is seems so natural, that it corresponds to the viewer's natural assumptions about where the gaze will be directed, or what the spatial and temporal relations are between shots, so that they don't notice the actual cuts. But this is, as I say, a figurative sense of "invisible," and describes the conceptual attitude of the viewer rather than the actual formal constitution or our real perceptual experience of the film. Even the most rudimentary analysis of a film will reveal the quite visible cuts that are used to create the otherwise coherent whole. A second look, often, is all it takes to see cuts where there seemed at first to be none.

Is this a sound basis upon which to base an account of "expression" in the cinema? If so, it would have to be extended to the digital era, and to account for digital editing. The question of expression, which is of course related to

67 Ibid., 58.

the question of artistic intention, which I have considered, but which must also be distinguished, as a separate problem, has indeed returned, as an issue in most accounts of digital editing, yet construed in a rather different sense. While Orpen's book was published in 2003, she does not mention what by then was the most significant new development in editing technology, namely the shift to non-linear, digital video editing, the consequences of which are thought by so many to be profound. The most significant change, many argue, or imply, is that while editing had not really been invisible in the past, it is now, or it can be, *truly* invisible. Does this mean, though, following Orpen, that the cinema has in fact lost its basic expressive capacities? This would seem to be the conclusion, or at least that the primary expressive technique, namely editing, which can now function in a dramatically more expansive fashion, to function primarily as a connective technique, but in a literally invisible manner, necessarily limits the possibility for expression, which, as construed by Orpen, requires that editing be seen, that its use be visible. In many of the apparent single-shot long sequences, or even single-shot films that I discussed in the previous chapter, the editing is often fully obscured, through digital means, obscuring as well any "expressive" aspect, reduced to the fact of editing. There are those who voice just such a concern, and who see in this the possibility of creating ever more believable, convincing or illusory, cinematic representations, which will more effectively fool viewers, seen less as intentionally designed artifacts and more as powerfully mimetic simulations. Yet, and at the same time, there are those who argue that it is with the radical expansion of the power of editing that the cinema has become wholly expressive, that it is now entirely subordinated to the intentional desires of the filmmaker. There is, it seems, a paradox here that needs to be resolved—derived, though, I am arguing, from the already murky concept of artistic expression.

Editing has become a more extensive technique, it is often argued, but whose presence, at the same time, can now be wholly obscured. In this respect, the changes to editing have seemed especially important. As we have seen, this is understood as a profound transformation. Lucy Fischer, for instance, argues that the advent of digital editing is as formally and aesthetically significant as the coming of sound in the 1920s and 1930s, requiring a transformation in our very conception of cinema.[68] The implication of her distinction, which she spells out only briefly in her short essay, is that such internal editing will not reveal itself to be such—what had been, and should be, visible, is now rendered wholly invisible. If, as Fischer suggests, much

68 Fischer, "Film Editing," 81.

editing is now effectively invisible, it is not clear whether an account of expressive editing like Orpen's may be sustained, and developed, or whether an alternative account of expression in the cinema is required, a wholly new theory of the cinema, as Fischer insists may be necessary. The question impinges on some of the most basic assumptions of film theory, assumptions that first emerged in the so-called classical era, before about 1960, but which have been sustained, as I argue, throughout the contemporary period, and which are being resurrected explicitly as theorists contending with the problem of the digital cinema return to some of the original proposals about the nature of cinematic representation and film art. As I have been suggesting, though, digital editing does not, *pace* Fischer, require a new theory or philosophy of cinema, but rather a clarification of some of the misunderstandings and conceptual confusions of much early film theory, many of which have been resurrected in recent accounts of digital cinema and digital editing, which will allow digital editing to be seen as consistent with the concept of the cinema as a technology, or better, a heterogeneous collection of many very different technologies, used in a variety of combinations, to various aesthetic ends, which is to say, as I will explain later, a set of possible tools for doing different things. As a new tool, digital editing systems can be understood as the sort of technological development that is wholly consistent with the history of the cinema as an art form, rather than, as many have argued, either foreign to an essentially cinematic aesthetic based in a putatively photographic identity, or "ontology," or as the source of a profound transformation of the cinema into a new kind of medium.

The questions about digital editing are in fact only the latest version of enduring questions about editing itself. While Orpen may be correct to say that it has been "taken for granted" that editing is expressive, in fact, throughout the history of film theory, the question has been whether its expressiveness itself is properly "cinematic" or not, or what sort of expressive role editing should properly assume, the answer to which has depended on how one has conceived of the basic identity of the cinema. The most famous contrast on this question, of course, is between the views of Sergei Eisenstein, and the other "formalist," montage theorists, who saw editing as the primary if not the sole source of meaning in the cinema, and André Bazin, and the other "realists" who preceded and followed him, who argue that its source was in the relatively autonomous process of photographic recording. The debate between the formalists and the realists, in other words, can be understood to be essentially a debate about editing, about montage, about its proper function and its aesthetic value. It is, then, in the digital era, another opportune juncture to reconsider the question of editing, of montage, when claims are being made again that it is either "everything" or "nothing," as Eisenstein had said about the debates in the 1930s.

Digital editing is being presented either as the comprehensive triumph of montage, or as having destroyed the integrity of montage as a feature of an exclusively photographic cinema. In order to engage with the question of, or the "problem" of, the digital cinema, in order to assess the claim that the advent of digital editing has had profound consequences on the very expressive possibilities of the medium, and as I have been arguing, it has been necessary to revisit the debates of film theory, and to assess the value of some of the most basic assumptions about the nature of cinematic representation. There is often in accounts of the digital cinema, and of digital editing, an explicit return to the basic assumptions of classical film theory. D. N. Rodowick, as he sets out to address what he sees as the fundamental change that has taken place since the incorporation of digital technologies into the apparatus of filmmaking, and the consequences specifically of digital editing, says: "Let's provisionally accept the premise of classical film theory that the photographic process is the basis of the film medium."[69] It is not clear to me, though, that even such a provisional acceptance is the best first step in an attempt to address the digital. I would propose that we subject such a premise to a thorough inspection, to determine whether it had ever been the case that one could have made such a claim in a convincing way. Certainly, there are many others whose opinions can be canvassed, from the past of film theory, who quite clearly do not accept such a premise.

The digital cinema is, however, commonly understood to have destroyed the cinema's basically photographic identity—for better or for worse—but it may well be that the advent of digital technologies, the ease with which they have in fact been incorporated into the filmmaking apparatus, reveals that the cinema was never so easily characterized as essentially photographic in the first place. Rodowick argues, like others, that the most significant effect of digital technologies is that they have transformed the cinema in such a way that it can no longer be distinguished from the other arts, that what had once clearly distinguished the cinema has been destroyed, so that the film image is now wholly "intentional," that the "causal" art of the cinema has been returned to the earlier condition of a "manual" art like painting. The cinema, he says, is no longer based fundamentally in a non-intentional "causal" process, which is how he conceives of photography. Worse, for Rodowick, as for so many others, is that much digital imagery in the cinema *seems* to be photographic, that is, it *seems* to be the product of an

69 D. N. Rodowick, *The Virtual Life of Film* (Cambridge, MA: Harvard University Press, 2007), 31.

autonomous, non-intentional, causal process of inscription or registration, when it is in fact only a "simulation," produced intentionally (and often, it is implied, or stated explicitly, with the specific intention to deceive). Computer generated images are, he says, "synthetic," unnatural composites, which do not clearly declare the fact of their composition—do not make their origin in "digital montage" visible to the viewer. To describe them as synthetic, though, suggests that this is a novel form of imagery, and that cinematic imagery has not traditionally been characterized by synthesis.

I don't think this is the case. If the imagery of the digital cinema is synthetic, then it is so in a sense that can be made continuous with the history of the cinema before computers became involved. In contrast to Rodowick, I do not see how there can be a clear distinction between a "causal" cinema, whose basis is photographic, and whose images are "created from physical reality," as he says, and an "intentional" cinema, whose basis is digital, and whose images are only simulations. I do not see how such a distinction of this kind can in fact be maintained. It is far more instructive, I think, to ponder the extent to which the "causal" photographic images have always been "synthetic" and put to some obviously "intentional" use in the cinema, even if such intentions may have been made deliberately obscure, and even if they are hidden, made "invisible," within the photographic image. Causality and intention are not so clearly opposite as Rodowick suggests, and it is characteristic of all technologies that the causal processes they embody can be harnessed to pursue any manner of intentional ends. In order to overcome the sort of false distinctions that are drawn between a supposedly "causal" technology like photography, and a supposedly "intentional" technology like digital imaging, it necessary to clarify the concept "technology," as I have already suggested, to understand what it means to say that the cinema is a technology, or a set of technologies, that can be employed to perform any number of possible functions.

Let's state the matter as simply as possible. Technology is, or technologies are, ways of doing things. This is, for instance the best way of thinking about photography—as a technology, which can, of course, be used in very many different ways, to different ends, and to various effects.[70] The

70 This is worked out in most careful form by Patrick Maynard, who argues that "photography" is "the name not of a class of things but of a kind of productive process, which frequently but not always issues in things of diverse use and interest." Maynard notes that "almost all writing about photography in our own times tends to begin with the alleged nature of the product rather than with its production and use. Photography itself," he says, "is taken as the productive process for producing photographs." See Patrick Maynard, *The Engine of Visualization: Thinking Through Photography* (Ithaca and London: Cornell University Press, 1997), 9.

advent of digital technology, and the theoretical concerns that it seems to have caused, reveal the need for the re-evaluation of the cinema in light of a more robust and informative account of technology, to describe the cinema *as* a technology, either in photographic or digital form. The fact of technology, and the fact of technological change, seem unavoidable when considering the art of the cinema. To what extent, though, and in what respect, is the cinema to be understood as a technological art? To what extent does technological change affect the cinema as an art? The emergence of digital editing, the transformation of the task of the editor from a basically mechanical and physical procedure—no more complicated than simply cutting and splicing pieces of film—to a complex and abstract process requiring elaborate and extensive computer systems, seems to have made this question a more urgent one. Something significant seems to have changed. What had once been the most "hands-on" aspect of filmmaking, the most tangible and material stage of film production, has now been incorporated into a comprehensively digitized process of image creation and manipulation, as the analog, photographic medium has been replaced by abstract and virtual systems of digital capture and computer manipulation and simulation.

It is not entirely obvious, though, *what* has changed, despite already considerable speculation about the advent of digital editing, and about the emergence generally of what is thought to be a wholly new digital cinema. Most of those who have considered the advent of digital editing insist at least that a change *has* occurred, while not always convincingly explaining the precise nature *of* the change. Most assume that the cinema is especially susceptible to the effects of technological development, and that such an obviously significant change, so extensive and far-reaching, must have profound, even radical, effects on the value of cinema itself. It is common to think of the cinema as a uniquely or essentially technological art, whose formal and aesthetic possibilities are necessarily altered by any change to its basic technology. "In the current climate of rapid technological change," as Rodowick says, "'film' as a photographic medium is disappearing as every element of cinema production is being replaced by digital technologies."[71] He is, it must be said, somewhat cautious in his claims, at least at the outset, suggesting at times that no clear break can be identified, while insisting, in a somewhat paradoxical manner, that a profound change has occurred nonetheless. He admits that "there is no inherent discontinuity cleaving the digital from the analogical arts," and states that, "[w]hile film disappears,

71 Rodowick, *Virtual Life of Film*, vii.

cinema persists".[72] Nevertheless, given that film's "ontology" was, as he says, "photographic," what persists as cinema is necessarily transformed as the "analogical" process of photography is replaced by "digital" systems of simulation. He argues that this specific change to the material, technical basis of the cinema has had a significant effect, that "the disappearance of film does matter and has had profound aesthetic and historical consequences".[73]

Rodowick's account is worth considering a final time, in some more detail, as an exemplary instance of the kind of claim made about the effects of digital technology on editing in the cinema, and insofar as he emphasizes the fact of editing in the change. Rodowick offers one of the most compelling versions of the claim that a profound change has taken place with the replacement of film by various digital processes as the main technical component of the cinematic apparatus. The most significant change, he argues, is to editing, or montage, long presumed to be the principal expressive technique in the cinema. Editing, though, understood indeed in some sense as an expressive technique, had been, or should be, Rodowick says, properly distinct from the more fundamental photographic process of image creation. Rodowick approvingly quotes Thomas Elsaesser, who says that "the digital image should be regarded as an expressive, rather than a reproductive medium."[74] As Rodowick then argues, emphasizing the transformation to photographic and cinematic representation, from a "reproductive" to a more "expressive" medium, "[t]he image becomes not only more painterly but also more imaginative."[75] The *causally* generated imagery of the cinema, created through automatic, non-intentional means, preceded the expressive manipulation of that imagery. Expression in the cinema, according to Rodowick, had been subject to the unique material limitations of photography, which is what has been overcome with the advent of digital editing. Expression had been necessarily constrained, he argues, insofar as editing was limited to the linear sequencing of images. The basic element of the cinema, the shot, consisted of an irreducible photographic image, the subsequent modification or manipulation of which was, if not impossible, was at least significantly constrained by the physical nature of the medium. Now, though, the images themselves may also be "edited," easily manipulated or modified after having been captured, a change that he insists has altered the very nature of the cinema.

72 Ibid.

73 Ibid.

74 Ibid., 105–6. See Thomas Elsaesser, "The New Hollywood: Cinema Beyond Distance and Proximity," in *Moving Images, Culture and the Mind*, ed. I. Bondebjerg (Luton: University of Luton Press, 2000), 192–93.

75 Rodowick, *Virtual Life of Film*, 106.

Rodowick describes what he sees as the radical discontinuity between the analog and the digital eras, as the traditional photographic process for creating the film image is replaced by what he calls the "digital event," that is, "any discrete alteration of image or sound data at whatever scale internal to the image. Elements may be added, subtracted, or refashioned interactively because the data components retain their separate, modular identities throughout the 'editing' process."[76] Editing is placed in quotation marks here, of course, because it has, Rodowick insists, undergone a profound transformation, and needs to be redefined. Editing as a term now designates a process that allows for the manipulation of all aspects of the cinematic image, at a fine, "modular" level. Editing had been limited in the past by what Rodowick effectively describes as the inherent material resistance of the photographic image, a supposedly unique kind of image that was understood to be effectively inviolable. Now, though, there is no such material resistance. The images in the cinema may be altered at will, altering the very nature of the medium. What we are presented with in the form of a "digital image," Rodowick argues, is very different from the traditional product of the camera, even though it may, to "ordinary perception," still look very much like a photograph. Digital images are characterized by what he calls "enhanced graphism," which has "rendered them more painterly; they are now more available to our creative intentions and less anchored to causal relations with the physical world."[77] In a word, more "expressive."

In the presence of photographs, he says, "we are inclined to judge by the criteria of perceptual realism, and as such to have faith that they are spatially similar to events we have witnessed and captured."[78] What we have forgotten, or what we are unable to see when we look at digital images, he says, especially since they can still look so much like photographs, is that "these 'outputs,'" as he calls them—the result of computer "transcoding" rather than photographic "transcription"—"may have no direct causal relationship to the events so witnessed or that the causal relation may be easily altered."[79] The "digital record," he says, in this respect, "is a symbolic form and thus, logically, is more similar to written description than to visual impression."[80] What he calls the photograph's "indexicality," using the term that has become so familiar, to mean direct causal connection

76 Ibid., 167.
77 Ibid., 123.
78 Ibid.
79 Ibid.
80 Ibid.

between subject and representation, is "seriously challenged" when the image is created with a digital camera, "and it is challenged all the more powerfully when we take advantage of the computer's powers to manipulate these records as information. For these reasons," he adds, "photographs and digital images provoke in us two very different kinds of ontological curiosity."[81] What Rodowick says he is trying to express is, "ontologically, the unbecoming of photography"[82]—such "unbecoming" manifested primarily in the expansion of the power of montage.

There is, though, and as I have suggested several times, a conceptual problem at the heart of such a claim. Rodowick's assumption is that there is a particular kind of thing that is the characteristic product of digital media, which is different in kind from the product of photographic media. Indeed, he suggests that what is now created digitally may no longer properly be called an "image," so profound are the differences between it and what is produced through what he calls the "analogical" process of photography. Rodowick assumes that there is an inherent quality that is characteristic of the process of photography, that photography is the means for producing a particular kind of thing, namely a photograph, that has a general property that can be found in all instances of photographic production. Rodowick is not alone in making such a claim. It is the most common claim made about photography, and it is the basis of most accounts of the advent of digital visual media, which are understood to produce a very different kind of thing—the *digital* image. Rather than emphasizing the many continuities between photography and digital imaging, the many shared features, and common functions, their formal and functional similarities as pictures, as depictions, as the bases for the creation of visual narrative fictions, and so on, Rodowick insists that they are fundamentally different phenomena, that require different theoretical analyses. Digital imaging and photography could, though, be understood as two forms of the same basic technology, as similar means for the same basic *productive process*. Stated as simply as possible, before I explore this in more detail below, this is a process for the inscription, or "marking," of photosensitive surfaces in order to create meaningful designs, which in turn can be put to a wide variety of possible uses. This, in the most basic sense, is what the technology of photography, in both analog and digital form, is used for. Whether that surface is chemical or electronic does not make a difference, at least in any general, or all-encompassing way. Difference emerges only through specific

81 Ibid., 124.
82 Ibid.

subsequent uses, not as a necessary result of a particular mode of initial production. Rodowick, though, like many others, insists that a particular process, photography, with a particular product, photographs, has been comprehensively replaced by another and fundamentally different process, described as "digital," which produces another kind of product, called a "digital image"—or a "simulation," or "digital event," or some other novel designation.

Editing, though, should not be conceived in terms of some special relation it is supposed to have to the photographic image, defined as a unique emanation of reality, or of the real world, and in stark contrast and resistant to the intentional manipulations of montage. Editing technology is one of the many technologies that are used, to various degrees, according to specific goals, and to perform various functions. Editing may be governed, we might say, by the pre-existing structures of reality, that is, it may be used to create realistic imagery—but this is not the same thing as saying that editing must be subordinated to the supposedly causal logic of photography, assumed somehow to be more intimately related to "reality." One function that editing may certainly perform is to construct a coherent representation of some preexisting event—to accurately depict an occurrence, whether real or fictional—by recreating a sense of temporal and spatial continuity. But this is by no means the only thing that can be achieved with editing. As Noël Carroll has said, quite simply, editing can be used to perform many different functions, some of which can become more dominant than others. "Paradigmatically," he says, in his account of "cinematic sequencing," the main function has been "to narrate a fiction. But there are other possible functions as well, including: to depict an environment, as in a travelogue or nature film; to make an argument, as in the case of *Fahrenheit 9/11*; to draw a comparison or contrast; to make a comment; to arouse an emotion; and so on."[83] These are only a few of the more common functions, among many possible and various functions. Such varied functions correspond to the many different things that photographs themselves are used for, and the many ways that photographs are used in the cinema.

Moreover, photography is not the only, nor necessarily the primary, technology of cinematic representation, and should not be understood to impose any particular constraint on, or regulate the use of, editing in any general or principled way. Many other technologies have been used to create markings on surfaces in order to engage cinema viewers imaginatively, including drawing, painting, scratching, and so on, such images, or "marks,"

83 Noël Carroll, *The Philosophy of Motion Pictures* (Malden, MA: Blackwell, 2008), 116.

available to be organized in various ways by means of editing. Increasingly, technologies of digital image creation are being used in ways that enhance and expand some editing capacities, while also relying on and reaffirming common editing techniques. If editing itself is thought of as a technology, as a way of doing things, many different and various things, then the transition from mainly photographic processes of image creation to digital processes would not be understood as having the kind of transformative effect on editing as it is commonly supposed to have. If one thinks of analog and digital processes not as the means for creating two very different kinds of things, but rather as different process for doing some similar things, and, of course, some different things, then no comprehensive claim can be made about the specific effects of digitization on editing in the cinema, such as the claim that the logic of "montage" has become any more dominant than any other "logic."

Rodowick's is perhaps the most sophisticated and detailed account of the supposed changes that have taken place with digitization, at the heart of which is a claim about the resulting change to editing, to montage, which he argues has indeed become dominant, as all aspects of the image become available for and susceptible to modification and manipulation, the effects of which had been limited in traditional film editing. In this respect, the digital cinema is understood as the triumph of intentional manipulation insofar as the materially intransigent nature of the photographic image has now been overcome, as photography has been fully incorporated into a semiotic realm, like any other "coded" form of representation. "In digital cinema," Rodowick says, "the image is always 'code,' that is, a combination of logically discrete elements completely open and available to changes in value, both perceptual and semantic."[84] Rodowick's book is a comprehensive summary, and further significant elaboration, of the sorts of arguments that have been made about the fate of the cinema in the digital era, presenting in perhaps the most powerful form the claim that there is a profound, perhaps essential, difference between the photograph and the digital image. Rodowick's argument is emblematic of a larger discourse, and a representative instance of a common argument in what is by now quite extensive literature on the consequences of digitization on the medium of photography and the art of cinema. The most significant consequence for Rodowick is the expansion of the power of montage, the enlarged capacities for the manipulation of the image that are provided by digital editing. As cinema becomes incorporated into an extensive apparatus of

84 Rodowick, *Virtual Life of Film*, 172.

digital media, he argues, it becomes governed by the logic of simulation and manipulation, of what he calls "transcoding," as opposed to photographic "transcription."

For Rodowick, although he does not attempt to qualify or clearly substantiate this claim with any comprehensive historical evidence, the specific material, technical aspects of the cinematic apparatus determine the aesthetic and formal possibilities of film art. As the photographic basis of the cinema has been replaced by digital processes, the cinema's possibilities are altered. Rodowick endeavors to explain the nature of the change that has taken place. "As digital processes come more and more to displace analogical one," he asks, "what is the potential import for a photographic ontology of film?"[85] It is not entirely obvious, he admits, that anything significant has changed, given that, for the most part, many "films" in the digital era still look like most films of the past, and still function according to familiar narrative and formal conventions. Despite what he calls the "many fascinating stylistic innovations" made possible by digital processes, these seem to have not substantially altered the look of most films, as "images," as pictorial depictions, as complex audio-visual representations. He does admit that "there is a strong sense in which what counts intuitively as an 'image' has changed very little for Western cultures for several centuries."[86] That traditional notion of an "image," he says, still seems to be the model for most digital compositions. The otherwise "synthetic" *digital* images "are still based on the same optical geometry as traditional cameras and rely on the same historically and culturally evolved mathematics of depth and light rendering descended from *perspectiva legittima*."[87] If this is the case, though, he asks, then what is it that has changed about image making in the digital era?

"Of course," he says, in answer to such a question, "what remains absent from the process of digital representation is what thinkers like André Bazin and Roland Barthes held fundamental to the photographic image: its causal force as a literal spatial and temporal molding of the originating event, preserved in a physical material."[88] The absence of such a "causal force," according to Rodowick, and many other theorists, makes the digital image very different from the photograph. They are, they insist, different kinds of things. The "digital image," Rodowick says, "partakes" of a

85 Ibid., 9.
86 Ibid., 11.
87 Ibid.
88 Ibid.

different "ontology." "Forged in the logic of information,"[89] it is based in systems of computation and numerical value, rather than a causal logic of registration or transcription. The digital image is thus radically different from photographs, and he insists that "a new philosophical perspective is required for these kinds of images, if they can still be called such."[90] Yet, as I tried to argue, it is not at all clear that a new philosophical perspective is necessary. The "logic of information," moreover, as a governing force in the realm of filmmaking, long predates the digital cinema, as I have suggested earlier.

***

Rodowick does occasionally acknowledge that photographs can perform different functions, and have different uses, but he is primarily concerned to identify a basic, common identity shared by all photographs. Instead of understanding digital imaging as a technological development in the history of photography, he insists that there is a fundamental difference between the kinds of things made photographically and those made digitally. He distinguishes these supposedly different kinds of images by describing their basic material constitution, their specific technical character, which determines what they are, what they can be used for, and what effects they can have. Such technical or mechanical determinants are what he calls, borrowing the term from Stanley Cavell, a medium's "automatisms." Rodowick distinguishes between those automatisms that are formal and conventional, that arise as a result of certain specific artistic practices and techniques, and those that derive from the physical, material basis of a particular medium. Yet it is to these latter kinds that Rodowick devotes most of his attention, without clearly explaining the differences between, say, the conventional use of perspective, or certain techniques of composition or color use, and the particular technical and mechanical affordances of a device like a camera, or of a medium like photography. And it is not at all clear why formal conventions characteristic of a particular art, the styles and genres that emerge over time, should be described as "automatisms," conceived, as the word inevitably suggests, as "automatic," with the implication that artists have little control over, or may make no judgments about, their use. To describe a genre, say, or a particular formal technique, as an "automatism," seems only to make more obscure the role that convention plays in the

89 Ibid., 175.
90 Ibid.

history of art. One could argue that some artists are less original than others, that they rely more on existing conventions, styles, and forms than others, that their use of such pre-existing conceptual and formal frameworks is "automatic," in a certain sense. Rather than this metaphorical, slightly pejorative use of the idea of the "automatic," though, Rodowick seems to be suggesting that all artistic activity is necessarily governed by some sort of "automatic" conventional constraint, or "limit." To describe conventions as "automatisms," though, is to obscure that fact that many artists are not at all constrained by the force of past convention, and that even apparently unoriginal, "conventional" artists often bring an individual awareness, even a quite explicit knowingness, to their use of conventions. Or it is perhaps better to say that convention is in fact best conceived in more positive, generative terms, not as a constraint, but rather as the basis of expression. "In encountering automatisms as limits," Rodowick says, "artists invent new creative strategies as ways of overcoming or transforming them."[91] But it is not clear how describing what he seems to mean as "convention" here as "automatism" contributes to an understanding of the role and force of the conventional. It gives the suggestion that an artist has little choice but to "overcome" what is automatically imposed on them, in a process of constant rediscovery. But that does not seem to describe a great deal of artistic work, which does not seek to overcome any particular conventional limitation, but rather employs conventions in a way that either reveals an understanding of them as conventions, and a willingness to use them in a more or less unaltered form, or manifests a determination to change them, and adapt them to new purposes. It is not clear what the re-description of the familiar term "convention" as an "automatism" reveals about the function of conventionality as a form of constraint, compared with or in relation to technical constraints, in the history of art.[92]

Rodowick does not really explore the question of the conventional as a term of historical enquiry or aesthetic analysis, though, focusing more on the

91 Rodowick, *Virtual Life of Film*, 44.

92 Jon Elster has offered a more subtle account of constraint and convention in art, but in other domains of human activity as well. As Elster says, the "restrictions imposed on or chosen by the artist can take many forms. A basic distinction is between *hard constraints* and soft constraints or *conventions*. Hard constraints are formal, material, technical, or financial restrictions on the selection and combination of units of the given medium. Conventions, as the word indicates, are restrictions that constitute a specific genre such as the sonnet of the classical symphony" (190). My argument above derives, although in rather undeveloped form, from Elster. Rodowick, I am suggesting, fails to make the sort of distinction that Elster does. See Jon Elster, *Ulysses Unbound: Studies in Rationality, Precommitment, and Restraint* (Cambridge: Cambridge University Press, 2000).

other kind of "automatism" that he describes, which seems more important as the originating source of the formal and conventional "automatisms." What he calls "mechanical or technical automatisms" play the central role in his argument, the built-in physical characteristics of a medium that effectively determine its aesthetic possibilities and the artistic value of the finished work. As Rodowick explains, "automatisms are both the material of aesthetic creation *and* the result of artistic practice. Automatisms in this sense are forms, conventions, or genres that arise creatively out of the existing materials and material conditions of artistic practice"[93]—they are, he says, adding another term, "elements," or "fundamental acts constituting the specific possibilities and necessities of a given medium."[94] While he offers a quite subtle account of the mutual relations between "material conditions" and "artistic practices," of the "elements" that are created through their combination, he quite clearly, if implicitly, establishes a hierarchy between the two, suggesting, in relatively traditional "materialist" terms, that practice is determined by material. "Why characterize these elements as 'automatisms'," he asks, raising the sort of question that I have posed above about the supposedly automatic nature of forms and conventions. Because, he suggests, his account of convention is derived directly from his materialist analysis of the determinant effects of any particular medium, of the necessary constraints that he assumes are imposed by a particular device or apparatus. Form and convention are merely reconceived as constraints that are comparable to, but are also importantly derived from, the basic material constraints imposed by any particular medium or device or apparatus. That media impose such constraints is most obvious in what he implies are the most constraining of media, namely photography and film. The "automatisms" of form and convention, he says, "are certainly inspired by the instrumental automatisms of photography and film, that is, the self-acting processes of mechanical reproduction."[95] Whether the "self-acting" nature of photography and film can legitimately be the basis for any informative definition of form or convention is not really explored in any detail by Rodowick, but it is clear that an artist's formal choices are, according to him, necessarily, and properly, constrained to the greatest degree by the autonomy of the photographic apparatus, and that these in turn are the bases upon which the supposedly more strict conventions governing photography and film are established. His argument is based on his specific understanding

93 Rodowick, *Virtual Life of Film*, 42.
94 Ibid.
95 Ibid.

of photography and film as somehow "self-acting" media, their mechanical and technical autonomy, it seems, necessarily constraining the artist's own autonomy to a greater extent than in any other art or other media—such constraint understood as the source of "convention" in all art, but of the most profound kind of technically generated convention that is supposed to be characteristic of photographic media.

"The instrumental qualities of cameras," he says, "the fact that they set into play a number of imaging processes at the request of, but also independently of, human action,"[96] is a central assumption of his argument, an assumption that he derives explicitly from André Bazin and Stanley Cavell. Rodowick rehearses the familiar arguments of the tradition, of which these two influential figures are so representative, according to which the photographic image is a unique medium, imposing a new kind of technical but also conventional and formal constraint on the artists, if not providing the basis for a wholly new kind of art, characterized by a clearer relation between material autonomy and conventional constraint. What an artist can do, according to this tradition, is determined and limited to an unprecedented degree by the "automatic" nature of camera's operations, which impose a new kind of constraint on the artist, subordinating their expressive intentions to the natural, or causal representations of the essentially autonomous photographic process of inscription. While Rodowick's is an especially powerful instance of this traditional argument, it is a widespread view among many film and media theorists of the past century.

96 Ibid., 43.

Conclusion

Abstract
In this concluding chapter, I summarize the overall argument of the book, and argue that rather than a "medium," the cinema should be understood as an amalgamation of various materials and technologies, supporting a wide range of techniques that can be used in whatever fashion chosen by the film artist. The advent of an aesthetics of digital montage, I argue, has precipitated a return to the basic debates of classical film theory, but for the most part many of the assumptions of those debates are simply reiterated, without advancing the theory of film, which must be able to accommodate even the most dramatic technological changes.

Keywords: Film editing, digital montage, technology, technique, art, aesthetics

Any account of "digital montage" will have to contend with the history of debates about the value of editing as a cinematic technique. There are, on the whole, and as we have seen, two basic views on the matter. Editing is either the primary and ultimate source of meaning in the cinema, or it is a secondary technique, subordinated to the more fundamental process of photographic recording. Among the most passionate advocates of the latter view was Andrei Tarkovsky, who deliberately contrasted his filmmaking style with his Soviet predecessors, creating some of the most elaborate and complex long-takes in the history of the cinema. In his theoretical writing, Tarkovsky explicitly rejects the claim that editing is, as he puts it, "the main formative element of a film."[1] This is the view, he says, of the "protagonists of 'montage cinema,' following Kuleshov and Eisenstein," who, he says, assumed that "a film was made on the editing table."[2] Tarkovsky

1 Andrei Tarkovsky, *Sculpting in Time: Reflections on the Cinema*, trans. Kitty Hunter-Blair (Austin: University of Texas Press, 1986), 114.
2 Ibid.

Furstenau, M., *The Aesthetics of Digital Montage: Film Editing and Technological Change*.
Taylor & Francis Group, 2024
DOI 10.5117/9789463722803_CON

insists that such an assumption is based on a mistaken understanding of cinematic art. "The idea of 'montage cinema'—that editing brings together two concepts and thus engenders a new, third one—seems to me," he says, "to be incompatible with the nature of cinema. ... The image is tied to the concrete and the material. ... The poetics of cinema, a mixture of the basest material substances such as we tread every day, is resistant to symbolism."[3]

Tarkovsky insists on the secondary and dependent nature of editing. "Editing," he says, "is ultimately no more than the ideal variant of the assembly of the shots, necessarily contained within the material that has been put onto the roll of film. ... Editing does not engender, or recreate, a new quality; it brings out a quality already inherent in the frames that it joins."[4] In his dismissal of the theory of montage, though, he offers a helpful observation, which may appear to contradict his account of editing as necessarily subordinate to the initial act of recording. A film, he says, is a complex whole, the result of the many techniques and devices that are available to the filmmaker. "No one component of a film can have any meaning in isolation: *it is the film that is the work of art*. And we can only talk about its components rather arbitrarily, dividing it up artificially for the sake of theoretical discussion."[5]

It is just for the sake of theoretical discussion that I have been distinguishing editing from the other elements of a film, dividing it up artificially, but with the acknowledgement, as Tarkovsky rightly says, that this is an arbitrary division. Indeed, I have been arguing that editing cannot be discussed except in relation to the many and various aspects of the cinema as a complex technical, formal and aesthetic enterprise. It is only as part of a greater technological assemblage, which historically has been in a constant state of transformation, that the emergence of digital editing can be understood. This may seem surprising since it has become common to describe the cinema as essentially photographic, and to see the emergence of techniques for the digital creation and manipulation of moving imagery as a threat to the medium's supposed essence. Given the cinema's complexity, the temptation to reduce it to something more manageable, to try to discover what is essential as opposed to what it is merely contingent, in order to account for and evaluate technological change in the apparatus of filmmaking, is understandable. Like the advent of synchronized sound, and other technical developments, including color, widescreen processes, and so on,

3 Ibid., 115.
4 Ibid., 116, 119.
5 Ibid., 114.

the introduction of digital and computer technologies into the cinematic apparatus has provoked what are often essentialist simplifications which see it as either anathema to the cinema as it had existed, or as the basis for an entirely new kind of cinema.

As we have seen, filmmakers and editors themselves, in contrast to the familiar accounts of film theory, often understand technologies as means to an end, rather than as constraints on or as autonomous forces channeling and determining their creative ambitions. The cinema, moreover, as I have argued, is not itself a medium, strictly speaking, but is best understood as an endeavor, or as an "environment," the term proposed by Thomas Ohanian, within which a variety of media may be employed in various combinations, in order to achieve any number of goals (to tell stories, to represent events, to forward arguments, and so on) and to do so in any number of possible styles (realistically, abstractly, through montage, or with long-takes, narratively, non-narratively, and so on), using photography, but also animation, painting, stop-motion, etc., and now also digitally synthesized or computer-generated imagery. The cinema, that is, cannot be reduced to any particular technical basis as the source of aesthetic value—especially since that basis is always, and at any moment, subject to change—but is the sum of the activities and choices of filmmakers, who pursue their aesthetic goals with the technical means that are available to them. Technological change is often the result of commercial, financial, or industrial decisions, either independently of or with regard for the aesthetic consequences. Inevitably, though, artists will discover opportunities, or new aesthetic and formal challenges, that lead to new modes of expression and representation, or, as is more common, they will use the new tools to accomplish their aesthetic goals according to traditional, prevailing conventions.

The fact of technology, though, seems more significant in an art like the cinema, and changes to technology more significant. Francis Sparshott long ago made such a claim, presenting it in perhaps starkest form. "More than any other art," he said, "film is technologically determined."[6] As a result, he said, "The history of film is the history of the invention of its means."[7] We could revise this, though, to say that the history of *all* the arts is the history of the invention and *reinvention* of their means. And we might add that those means, those technical, material devices that emerge and are put to use by artists, are precisely, and only, *means*. Artists

6 F. E. Sparshott, "Basic Film Aesthetics," *Journal of Aesthetic Education* 5, no. 2 (April 1971): 12.

7 Ibid.

employ the technical means available to them in order to realize, we might say, a non-technical, aesthetic goal. Among filmmakers, and film editors, there is on the whole a considerable willingness, if not often a distinct enthusiasm, for digital technologies, but only insofar as they are seen as viable means for either the traditional aesthetic goals they have sought to achieve, or the basis for imagining new goals. This is not to say that, as we have seen, and as in any other complex human endeavor, there are not debates, disagreements, and deep concerns among filmmakers about new technologies in the cinema, or that there is anything like universal acceptance of the new digital technologies. But there is considerable evidence that, on the whole, filmmakers—and by this phrase I mean all those involved in substantial ways in the aesthetic undertaking of filmmaking—have welcomed such technology. While technological change has aesthetic and artistic consequences, they are not typically seen as fundamental. Whatever tools or materials are available to artists, they will find ways of achieving their artistic goals. This could be generalized, to say that all artistic goals are intellectual, conceptual, "transcending," as Gérard Genette argues, their technical "immanence," as part of a complex, intentional artistic action. "The modes (a provisional term) of immanence," Genette says, "have heretofore been treated as modes of existence *simpliciter*."[8] The immanent or technical basis of a work or art is just that, a basis, with which an artist performs an aesthetic function, creating an aesthetic relation, through a process of transcendence, or overcoming. The technical aspect of the efforts to achieve the goal are guided by the intentions of the artist, by their specific ambitions—which may, of course, be altered, changed, revised, in the face of the technical challenges that they confront, or opportunities they encounter, expected or unexpected, having chosen particular tools. Such a view is implicit in the remarks by the film editor Craig McKay, interviewed by Vincent LoBrutto. "I think less about the mechanics of filmmaking," says McKay. "I'm not thinking about the Moviola, or the synchronizer, or the track. To me it's an intellectual, emotional experience."[9]

The cinema, I have been arguing, is best understood as an endeavor, using various media in some combination, rather than as itself a singular much less unique medium. It is, I have suggested, best thought of as an "environment," within which various media may be deployed in various

8 Gérard Genette, *The Work of Art: Immanence and Transcendence*, trans. G. M. Goshgarian (Ithaca: Cornell University Press, 1997), 11.

9 Vincent LoBrutto, *Selected Takes: Film Editors on Editing* (New York and Westport, CT: Praeger, 1991), 205.

combinations. Many film theorists, though, tend to reduce this complex sphere of activity to a single medium, the photographic image, which they insist is the most fundamental, the most "cinematic." While there are many differences among film theorists, most accept the photographic assumption. The debate tends to be whether the photographic origin of the cinema is to be acknowledged as regulating and delimiting the aesthetic possibilities of filmic representation, or if it is to be overcome through the deliberate imposition of a technique for the intentional modification of such causally produced "raw material," the most significant technique being editing. The two most common claims about the cinema, as we have seen, are that it is essentially the art of the camera—that its meaning is created at the moment of recording, that it is derived from a kind of fidelity to, or respect for, the integrity of the recorded image, and that all other techniques are necessarily subordinate to the cinema's basically photographic identity—or that it is the art of editing—that meaning is the result of the arrangement and the assembly of images or sequences of images, of, that is, the necessary and inevitable manipulation of the material that is gathered up by the camera, which to that extent is *nothing more* than a recording device, awaiting the subsequent and fateful interventions of the filmmaker and editor.

The cinema is characterized by a technical complexity, which often tends to be reduced to a simple claim about the essential or definitive quality of any particular technical aspect or specific material, to suggest an ultimate unity against complexity. In a contribution to a symposium on film and technology, Peter Wollen argued some time ago that "the technology of the cinema is not a unified whole, but is extremely heterogeneous. It covers developments in the fields of mechanics, optics, chemistry, and electronics. It covers the Latham loop, Scotch tape, the densitometer, the zoom lens, magnetic tape, and so on and on (the list is long)."[10] Long, and, I would add, never complete and always changing. It may now include, of course, the computer, software, the internet, algorithms, and all the various aspects of digital technology. In fact, Wollen seems to anticipate the digital future of the cinema, the origins of which can be traced back to the introduction of electronic technologies, first in sound recording. "It is with sound," he says, "that the truly modern technology of electronics first made a real impact on the cinema."[11] The synchronization and amplification of sound became possible with the invention of the audion tube, which was the crucial

10 Peter Wollen, "Cinema and Technology: A Historical Overview," in *The Cinematic Apparatus*, eds. Teresa De Lauretis and Stephen Heath (London: Macmillan, 1980), 20.
11 Ibid., 16.

component for large and effective loudspeakers and public address systems, and with the photoelectric cell, which made the optical soundtrack feasible as a component of the filmstrip itself, where sound could be registered at the same time as the image. What had been achieved for sound would, he predicted, soon become possible for imagery as well.

Recorded electronically, sound could of course be much more easily manipulated and edited than the image, but it also meant, given "the need for sound-image synchronisation," that "every aspect of timing had to be standardised."[12] As Wollen explains, while sound introduced new possibilities, it "also introduced new obstacles and a chain of bizarre secondary problems and solutions,"[13] as when, for instance, the loud carbon lights that had been standard were replaced by quieter tungsten lighting, which in turn required more sensitive panchromatic film stock, which was more complex technically to process and correct, requiring even more standardization. All of this (and many other factors) meant that the laboratory processes had to become much stricter and more technically precise, with the consequence, as he says, that "the laboratory became completely divorced from the work of the director and the cinematographer; it became an automated industrial process with its own standard operating procedures."[14] The addition of color added further complexity, with the effect that the "opacity and autonomy of the laboratory was further accentuated."[15] What had (inadvertently, and unexpectedly) created this problem, though, namely the introduction of electronic sound recording, might, Wollen predicts, eventually emerge again as the solution. "It might be added," he says, as he imagines the further stages in this dialectical process of technological development, "that one of the attractions of colour video, which uses electronic rather than chemical technology for colour, is that there is no laboratory involved. ... It is only a matter of time," he suggests, "before electronic technology gains the ascendency in image as well as sound."[16]

This, I argue, is just what has been realized in the so-called digital cinema, which can be described as the comprehensive application of electronic technologies to processes that had been largely chemical and mechanical, requiring significant divisions of labor. The advent of the digital is best understood as a stage in the process of technological development as Wollen

12 Ibid., 17.
13 Ibid.
14 Ibid.
15 Ibid., 19.
16 Ibid., 19.

describes it, which consists of a constant drive for improvement, for solutions to particular technical and formal problems of filmmaking—to which Wollen was attuned, as himself a filmmaker—with the almost inevitable creation of further unforeseen problems, that in turn require yet another solution. The cinema is best understood in the context of what the historian of technology Robert Friedel has called a "culture of improvement," by which he means "the ascendancy of values and beliefs permeating all levels of society that 'things could be done better.' This is not," though, he is very keen to insist, "the same as a faith in progress or the belief that the necessary trajectory of history or human experience was upward, but rather the simpler daily understanding that human beings have the capacity to improve how they do things."[17] What this also means, though, as Wollen likewise implies, is that humans typically fail to consider or predict the consequences that specific technical improvements might engender. In this respect, improvement may well be the opposite of progress. As Friedel says, improvement is contingent. This means, first of all, that there "is no absolute or universal standard for what constitutes an improvement; an improvement is determined by individuals (who may or may not work in concert with others) with specific goals at specific times. What may be an improvement in carrying out a task or making something at one time for one person may be no such thing for anyone else—or even for that same person at another time."[18] Improvement understood as contingent, he says, "directs our attention to the immediate technical motivations for innovation."[19] While this may seem obvious, he admits, it is typically overlooked in most accounts of technology, which tend to take the form of what he calls "post hoc kinds of explanations—that is, explanations that account for the emergence of a technology based on the final effects that the technology has. Looking at improvement as an act whose character is completely contingent on the immediate goals of the actors," he says, "gets us away from this sort of thinking."[20]

A good many accounts of digital cinema, and digital editing, also take the form of post hoc explanations—usually as claims that computer technology has changed the very nature or identity of the cinema. As an example of the sort of complex context within which manifold and often

17 Robert Friedel, *A Culture of Improvement: Technology and the Western Millennium* (Cambridge, MA and London: MIT Press, 2007), 2.

18 Ibid., 3.

19 Ibid.

20 Ibid.

competing technical activities are undertaken, though, the cinema is best described as a specific community of actors (and, of course, of actual "actors," or thespians) who are for the most part engaged in various efforts that routinely become understood as susceptible to improvement. To offer, in preliminary fashion, only one well-known example, there is the famous and fruitful collaboration between Orson Welles and Gregg Toland to develop the specific lenses necessary to achieve the sort of deep-focus shots that Welles wanted for *Citizen Kane*. While this is a perhaps notable example, it is by no means unique. Filmmaking is an almost inevitably ad hoc enterprise, improvisatory, a response to a series of formal problems requiring technical solutions, all undertaken in the context of massive financial investment and commercial interests, in order to satisfy the demands of a mass audience, whose discriminating tastes are far more subtle and variable than critics of "mass" culture have long suggested.[21] As Jean Mitry pointed out, with his usual candor, "one does not make a film to make a film; one does it to *make money*."[22] His point, with which he begins his analysis, is that given its enormous complexity as a technical undertaking, filmmaking has to be based on as sound a financial model as possible, in a more or less comprehensive process of industrialization. It is just because of this, he argues, that production in the cinema "has reached a level of average quality which guarantees, to some extent, the value of the spectacle. If there is a plethora of mediocre films," he says, "their mediocrity is most often found in their content rather than their form, and very bad (that is, badly made) films are becoming increasingly rare."[23]

Mitry was writing in 1963, but his observations are perhaps even more relevant now—for better or for worse, the cinema is operating, I would venture to say, in almost all its aspects, at a level of above-average quality. Technically, we have reached a basic standard of considerable proficiency, and there is a large class of filmmaking professionals, skilled artisans and technicians, but also highly trained actors and musicians, and many others, who make possible the quite consistent production of very good, in the strict sense of well-made, technically proficient, movies. Whether they are "good," in a more evaluative, aesthetic sense, is another question. While technically well-made, very many films (most?), in terms of their content,

21 "Philosophical aesthetics in the twentieth century," write Noël Carroll, "has shown a striking inability to come to terms with mass art." See Noël Carroll, *A Philosophy of Mass Art* (Oxford: Oxford University Press, 1998), 15.

22 Jean Mitry, *The Aesthetics and Psychology of the Cinema*, trans. Christopher King (Bloomington and Indianapolis: Indiana University Press, 1997), 7.

23 Ibid., 5.

are at best mediocre—much is simply terrible. Artistic accomplishment, though, is not determined, made any more or less likely, by any particular material or technical basis. The establishment, though, of a firm technical, artisanal, and financial foundation, which may accommodate even the most significant technological changes, makes possible the regular production of technically highly proficient films that are also considerable aesthetic achievements. It is best to think of such an endeavor, the collective endeavor that is the cinema, as an instance of effects of constant efforts at improvement, which, as Friedel insists, "refers to the perceived ends of the actors" engaged in such an activity, "essentially short-term and local, not to more cosmic outcomes."[24] Friedel continues:

> While social and material progress ... may indeed be loudly touted as both a goal and an achievement, this does not so much describe immediate actions as it attempts to link these actions to ostensibly larger and worthier aims, indeed aims implicitly sanctioned by God and Nature. To talk of progress is to speak in essentially teleological terms, to suggest that technology (or whatever is the subject of the term) is moving linearly toward a divine end. The reader is free to believe this or not; it is essentially irrelevant to the way the story of technology is laid out here.[25]

Most film theorists, of course, cannot be accused of a blind faith in progress, or of thinking that the cinema's success as an art is sanctioned by God. There is, though, often an implicitly teleological assumption in a great deal of theoretical analysis of cinema, if in fact a kind of negative teleology, a fear of decline and dissolution, of substitution and deterioration. Insofar as a particular aspect of the technological apparatus is elevated above any other, in a formal hierarchy, there is an assumption about the essential aesthetic value of the cinema, and a tendency to conclude that the replacement of one element with another is what will *prevent* the cinema from achieving its presumed inherent purpose. As Stanley Cavell says, for example, in perhaps his most famous formulation, "The idea of and the wish for the world recreated in its own image was satisfied *at last* by cinema. Bazin calls it the myth of total cinema. But it had always been one of the myths of art."[26] On the face of it, this is perhaps an unobjectionable claim, but it

24 Friedel, *Culture of Improvement*, 10. He means, of course social "actors," not thespians.
25 Ibid.
26 Stanley Cavell, *The World Viewed: Reflections on the Ontology of Film* (New York: Viking, 1971), 39; emphasis in original.

assumes that the cinema's development is guided by something like what Friedel characterizes as a "cosmic" wish, that the technical development of the cinema is governed by an essentially progressive teleology, achieving its putative goal, of recreating the world "in its own image," *at last*. This may well be something that Cavell looks for in art. But it is an instance of a post hoc explanation, which disregards the fact that there are an endless number of things that one may seek to do in cinema, and to have the world "recreated in its own image" may be one, but it is by no means the single goal towards which cinematic art necessarily tends, or in which it may be thwarted by technological change and material transformation.

Much of this is posed in film theory in the language of ontology. Influenced by Cavell, D. N. Rodowick has sought to answer the question whether the cinema in the digital era is still capable of performing what Cavell had identified as its "ontological" function. His affiliation with Cavell's account of photographic ontology is explicit. For Cavell, he says, film represents a fundamental fact of the human condition, or as Rodowick puts it, "the phenomenology of film projection expresses for Cavell both the metaphysical isolation of the modern subject and its possible overcoming. Film presents to us not only the visible world or the world as visible, but also our conditions of viewing in just this way. ... Photography and film mechanically reproduce the subjective conditions and paradoxes of skepticism in the form of a possible philosophical solution."[27] This is, as Rodowick presents it, an account of film's specificity as a medium, which presumably is manifested regardless of the content of any particular film. He acknowledges that there is, in the work of philosophers like Noël Carroll, "a good case against medium-specificity arguments, or the idea that there is something ontologically unique about photographic and cinematic images," and he admits that, really, "what we call 'film' is an unstable or variable object unanchored by familiar aesthetic criteria of self-identity."[28]

Nevertheless, he says, and while Carroll and others seem to have definitively dispensed with what had been the question guiding classical film theory, namely the question of film's identity, its specificity as a medium, Rodowick "confesses" that, "I have not yet convinced myself that the question is closed. At the beginning of the twenty-first century," he says, "we have come full-circle again. And as the photographic medium threatens to disappear forever on the scrapheap of technology, suddenly the main

27 D. N. Rodowick, *The Virtual Life of Film* (Cambridge, MA: Harvard University Press, 2007), 69.
28 Ibid., 30.

questions of classical film theory seem worth revisiting."[29] This is, indeed, what I have been endeavoring to do in this book, but finding, I think, quite different answers. All of this returns Rodowick, again, to questions of editing in the digital era, which has become what he calls a process of "synthesis," transforming and expanding the powers of editing, or what Rodowick calls "digital image synthesis," in such a way that is supposed to radically alter the cinema itself. Rodowick insists that "the emergence of digital filmmaking and, more radically, of digital image synthesis might mean that the very nature of the medium is changing—in sort, becoming something that is no longer *film* in the ordinary sense of the term."[30] He proposes to offer, as he says, "some new philosophical answers to some old theoretical questions. What is a medium? Is 'film' in the conventional sense fundamentally tied to photography? When filmmaking and viewing become fully digital arts, will a certain experience of cinema be irretrievably lost?"[31] In light of the changes that he sees occurring at the very foundation of the medium of film, Rodowick offers a slightly grammatically altered version of the central question of classical film theory: "What was cinema?"[32]

While in the past tense, this is still an instance of the kind of argument that I have been resisting, in the form of what Ian Jarvie has called a "what is" question. Arguments about the nature of film, attempts to determine what it 'is,' he says, "highlight the conservative aspect of essentialism, which essentialists are often blind to. 'What is film?' leads to various kinds of wild goose chases, but the most insidious is the idea that film has a fixed nature and that we do not 'know' what film is until we have pinpointed that fixed nature, whether directly by definition, or indirectly by approximating differences from other things."[33] Peter Wollen had been sensitive to the dangers of "what is" questions, and offers the model of heterogeneity as an alternative to the more common tendency to identify a singular cinematic identity. "In the past," he says, writing in 1978, "theorists have tended to stress and even essentialise one or other area of technology at the expense of others," so that "the heterogeneity of the cinema is reduced to one subset of determinations in a reductionist manner."[34] In particular, he says, "too much attention is usually paid to the recording stage at the expense of the

29 Ibid., 30–31.
30 Ibid., 31.
31 Ibid.
32 Ibid., 25.
33 Ian Jarvie, *Philosophy of the Film: Epistemology, Ontology, Aesthetics* (New York and London: Routledge and Kegan Paul, 1987), 157.
34 Wollen, "Cinema and Technology," 20–21.

laboratory and the theatre."[35] As a result, he argues, little is known about various technical aspects of film technology that are in fact as important, if not more so, than the camera. "The history of the technique of editing," he says, "for instance, is rather neglected."[36]

Since Wollen's observation, there has been a considerable amount of very important historical work on the origins and the development of editing in the cinema, which has transformed much thinking on the subject. Yet as a specific formative element, editing is still often overlooked, or misunderstood. When it is considered, it is still often perceived as a threat to the integrity of the original camera image, as a formative technique that should be properly subordinated to the initial process of recording. Or, according to the other view on the matter, it is the technique that overcomes the limits of a simple recording, and that creates new meanings. In either case, though, the source of meaning in the cinema is thought to be the result ultimately of the tension between recording and assemblage. Expression in the cinema is thought either to be attenuated by the unique quality of the image created by the process of photographic recording, understood as causal and, in many important respects, non-intentional, or it is the result of the effort to impose on such recordings, mainly through the techniques of editing, an intentionally generated meaning. As I have been suggesting, though, there is a way out of this impasse, which must involve, though, a deliberate effort to return to the "neglected" topic of art of editing, understood now as the aesthetics of montage.

***

The history of film theory, one might say, has been structured by this basic debate about the limits or capacities of editing (and other formal techniques of manipulation and alteration) in relation to the photographic image understood as the basic medium of the cinema. The familiar distinction between formalism and realism in film theory is essentially a contrast between two views on the question of the proper attitude towards the photographic, the difference being how one thinks that fact should be accounted for, whether, as André Bazin famously put it, one should put one's faith in the image or in reality, in, that is, the deliberate, formal manipulation of the material gathered by the camera, or in the careful preservation of the traces of the real that have been captured photographically. Irving Singer, in *Reality*

35 Ibid., 14.
36 Ibid.

Transformed, is one of the few who have tried to resolve and overcome this basic distinction, to resolve the disagreements between what he calls the two types of classical film theorist, the realist and the formalist. The realists, he says, "emphasize that film records properties of the physical world that lend themselves to the photographic process," while the formalists "call attention to the technical means by which a filmmaker goes beyond the real world in order to express his or her artistic vision."[37] As a first step towards resolving this difference, Singer makes a simple but powerful point, that film is a medium of communication. Understood as such, we may, he argues, overcome, or, as he says, "harmonize," the views of realism and formalism, which are thought to be irreconcilable. Rather than two distinct views, which one must choose between, Singer argues that they may in fact be unified by a "theory of transformation," which he endeavors to develop, which is in effect a theory of expression or of cinematic communication. "In being a vehicle of communication," he says, "film changes reality through techniques that matter because of meanings they are able to generate. ... Films transform reality by exploring imaginative possibilities of communication. ... The communication present in cinematic art," he says, "originates with a filmmaker through technological devices that are suitable for conveying whatever ideas and feelings he or she wishes to express."[38]

The elaboration of a robust theory of communication, or expression, Singer argues, should be the main goal of film theory, or a philosophical analysis of the cinema, and will be the basis for what he saw as the still unresolved debate between formalists and realists. Rather than "communication," though, or "expression," which I have been suggesting are themselves fraught concepts in the theory of art, it is better to think in more expansive terms about intention, or intentionality. In any case, a resolution of the supposed differences between the realists and the formalists of classical film theory would be a step towards such a goal. The debates between classical film theorists had, for the most part, simply been dismissed, "swept aside," as Singer says, with the advent, after the 1960s, of new modes of film theory, influenced by new developments in linguistics, semiotics, and psychoanalytic and ideological critique. Such "Grand Theory," as he calls it, following David Bordwell and Noël Carroll—or, rather, he says, what were in fact several rival grand "Theories"—failed for the most part to realize their ambitions. In fact, he says, they were unable to get beyond the fundamental

37 Irving Singer, *Reality Transformed: Film as Meaning and Technique* (Cambridge, MA and London: MIT Press, 1998), 1.
38 Ibid., 6–7.

question that had been at the heart of the debates of classical film theory: "Turning away from the quarrel between realism and formalism, the Theories generally incorporated some version of either one or the other. But it was usually presented with little clarity and less than sufficient concern about alternate views equally worth considering."[39]

It is the quarrel between realism and formalism, the classic debates about the value of photography as opposed to the value of editing, that has in fact been resurrected in recent film theory, as it has contended with the changes that have taken place with the incorporation of digital technologies into the apparatus of filmmaking, and specifically the advent of digital editing. It has been my argument in this book, though, that such a debate is based on a set of mistaken assumptions about the role that various media and techniques play, and the changing value that they have had, in the history of the cinema as a complex and collective endeavor. Editing is not a particular component of filmmaking, the general value of which can be determined in any definitive way, but only one of the many, and always changing, elements of the heterogeneous cinematic apparatus. When editing is considered, it is often presented as the source of cinematic meaning and artistry, which has given meaning to otherwise inchoate photographic imagery, or it is understood to be properly constrained by and undertaken according to the original and more significant process of photographic recording. In generous accounts, it is understood as what allows sense to be made of the even more fundamental products of the original recording process—the photographic image as the source of the proper limiting constraints on editing. Digital editing, though, reveals that this is a mistaken view, and that no such straightforward hierarchy of means can be established. The apparently greater freedom that digital editing has engendered, what I am calling an aesthetics of digital montage, for better or for worse, depending on one's theoretical proclivities, must in fact be seen to be continuous with the freedom that has always pertained, and the limits that have been imposed, that have characterized filmmaking as soon as it was conceived of as an artistic undertaking.

39 Ibid., 4. See David Bordwell and Noël Carroll, eds., *Post-Theory: Reconstructing Film Studies* (Madison: University of Wisconsin Press, 1996).

Acknowledgements

I was first asked to write a book on digital editing in the cinema (far too long ago now!) by André Gaudreault, who provided material and intellectual support to me as a member of his research group *Technès*, funded by the Social Sciences and Humanities Research Council of Canada (SSHRC). His enthusiasm, scholarly care, and theoretical curiosity are an inspiration. I would also like to thank the other member of the *Technès* team, at the University of Montreal, especially Santiago Hidalgo and Marnie Mariscalchi. I first wrote about digital editing with Martin Lefebvre, who has guided me in much of my thinking on the subject in the many conversations we have had about film and theory. I want to thank David Richler, who as a doctoral student in Cultural Mediations at Carleton University assisted me in some of the original research. Other students have contributed to my thinking, in the many conversations that I have had in seminar rooms and at conferences, if not about film editing or digital montage specifically, but the theory and philosophy of film art more generally, most notably Owen Lyons, Matthew Croombs, Murray Leeder, Dino Koutras, and Kate Rennebohm. Many others have helped, in many ways, over the years, as I have passed various ideas by them, trying to make my thoughts clear, including colleagues, current and former, at Carleton, Aubrey Anable, Ira Wagman, André Loiselle, Laura Marks–but also Jerry White, Mike Baker, and Christine Straehle. The most basic ideas at the heart of my argument here were first worked out in my doctoral research, in the Graduate Program in Communications, at McGill University, with considerable guidance and support from Will Straw, but also with help from my fellow GPC students, Geoff Stahl, Anthony Kinik, Mark Hayward, and Peter Urquhart, who still provide friendship and intellectual support. Some parts of the argument in this book were presented at various conferences, including the annual meetings of the Film and Media Studies Association of Canada (FMSAC) and the Society for Cinema and Media Studies (SCMS).

Bibliography

Andrew, Dudley. *The Major Film Theories: An Introduction*. Oxford: Oxford University Press, 1976.

Andrew, Dudley. *What Cinema Is! Bazin's* Quest *and its Charge*. Malden, MA: Wiley-Blackwell, 2010.

Arnheim, Rudolf. *Film*. Translated by L. Marie Sieveking and Ian F. D. Morrow. London: Faber and Faber, 1933.

Arnheim, Rudolf. *Film as Art*. Berkeley: University of California Press, 1957.

Arnheim, Rudolf. *Film als Kunst*. Suhrkamp, 2002. First published 1932.

Aronson, Ian David. *DV Filmmaking: From Start to Finish*. Sebastopol, CA: O'Reilly Publishing, 2006.

Aumont, Jacques. *Montage Eisenstein*. Translated by Lee Hildreth, Constance Penley, and Andrew Ross. Bloomington and Indianapolis: Indiana University Press, 1987.

Barthes, Roland. *Camera Lucida*. Translated by Richard Howard. New York: Hill and Wang, 1981.

Bazin, André. *What Is Cinema?* Translated by Hugh Gray. Berkeley: University of California Press, 1967.

Bazin, André. *What Is Cinema?, Vol. II*. Translated by Hugh Gray. Berkeley: University of California Press, 1971.

Bazin, André. *Orson Welles: A Critical View*. Translated by Jonathan Rosenbaum. Los Angeles: University of California Press, 1991.

Bazin, André. *Qu'est-ce que le cinéma?* Paris: Editions du Cerf, 2001. First published 1958.

Bazin, André. "Editing Prohibited." In *What Is Cinema?*, translated by Timothy Barnard, 73–86. Montreal: Caboose, 2009.

Beardsley, Monroe. *Aesthetics: Problem in the Philosophy of Criticism*, 2nd edition. Indianapolis: Hackett, 1981.

Belton, John. *Widescreen Cinema* (Cambridge, MA: Harvard University Press, 1992).

Belton, John. "Digital Cinema: A False Revolution." *October* 100 (Spring 2002): 99–114.

Benjamin, Walter. "The Work of Art in the Age of Mechanical Reproduction." In *Illuminations*, edited by Hanna Arendt, translated by Harry Zohn, 217–51. New York: Schocken Books, 1969.

Bizony, Piers. *Digital Domain: The Leading Edge of Visual Effects*. London: Aurum Press, 2001.

Bordwell, David. *The Cinema of Eisenstein*. Cambridge, MA: Harvard University Press, 1993.

Bordwell, David and Noël Carroll, eds., *Post-Theory: Reconstructing Film Studies*. Madison: University of Wisconsin Press, 1996.

Bordwell, David. "Intensified Continuity Visual Style in Contemporary American Film," *Film Quarterly* 55, no. 3 (Spring 2002): 176–28.

Bottomore, Stephen. "Shots in the Dark: The Real Origins of Film Editing." In *Early Cinema: Space, Frame, Narrative*, edited by Thomas Elsaesser, 104–13. London: BFI, 1990.

Branigan, Edward and Warren Buckland, eds., *The Routledge Encyclopedia of Film Theory*. London and New York: Routledge, 2015.

Burch, Noël. *Theory of Film Practice*. Translated by Helen R. Lane. London: Secker and Warburg, 1973.

Campbell-Kelly, Martin and William Asprey. *Computer: A History of the Information Machine*. New York: Basic Books, 1996.

Carroll, Noël. "Toward a Theory of Film Editing," *Millennium Film Journal*, 3 (Winter/Spring 1979): 79–99.

Carroll, Noël. *Philosophical Problems of Classical Film Theory*. Princeton: Princeton University Press, 1988.

Carroll, Noël. *Theorizing the Moving Image*. Cambridge: Cambridge University Press, 1996.

Carroll, Noël. *A Philosophy of Mass Art*. Oxford: Oxford University Press, 1998.

Carroll, Noël. "Film as Art: Introduction." In *Philosophy of Film and Motion Pictures: An Anthology*, eds. Noël Carroll and Jinhee Choi, 7–18. Malden, MA and Oxford: Blackwell, 2006.

Carroll, Noël. *The Philosophy of Motion Pictures*. Malden, MA: Blackwell, 2008.

Cavell, Stanley. *The World Viewed: Reflections on the Ontology of Film*. New York: Viking, 1971.

Ceruzzi, Paul. *A History of Modern Computing*. Cambridge, MA: MIT Press, 1999.

Cinemetrics. "Info." https://cinemetrics.site/?selection=,,,,,#/info.

Cubitt, Sean. *The Cinema Effect*. Cambridge, MA: MIT Press, 2004.

Currie, Gregory. *An Ontology of Art*. London: Macmillan Press, 1989.

Cutting, James E., Jordan. E. DeLong, and Christine E. Nothelfer. "Attention and the Evolution of the Hollywood Film," *Psychological Science* 21, no. 3 (2010): 432–39.

Dalle Vacche, Angela. *André Bazin's Film Theory: Art, Science, Religion*. Oxford: Oxford University Press, 2020.

Dancyger, Ken. *The Technique of Film and Video Editing: History, Theory, Practice*, 5th edition. Burlington, MA: Elsevier/Focal Press, 2011.

Daney, Serge. *L'exercice a été profitable, monsieur*. Paris: POL, 1993.

Danto, Arthur. *The Transfiguration of the Commonplace*. Cambridge, MA: Harvard University Press, 2005.

Danto, Arthur. "Deep Interpretation." In *The Philosophical Disenfranchisement of Art*, 47–67. New York: Columbia University Press, 1983.

Davies, David. *Art as Performance*. Malden, MA: Blackwell, 2004.

Davies, David. "Digital Technology, Indexicality, and the Cinema," *Rivista di estetica* 46 (2011): 45–60.

Davies, Stephen. "Ontology of Art." In *The Oxford Handbook of Aesthetics*, edited by Jerrold Levinson, 155–80. Oxford: Oxford University Press, 2003.

Dayan, Daniel. "The Tutor-Code of Classical Cinema," *Film Quarterly* 28, no. 1 (Autumn 1974): 22–31.

De Lauretis, Teresa and Stephen Heath, eds. *The Cinematic Apparatus*. London: Macmillan, 1980.

Dewey, John. *Art as Experience*. New York: Penguin, 2005. First published in 1934.

Dhir, Meraj. "The Modern Entertainment Marketplace, 2000–Present: Editing." In *Editing and Special/Visual Effects*, edited by Charlie Keil and Kristen Whissel, 156–71. New Brunswick, NJ: Rutgers University Press, 2016.

Dixon, Wheeler Winston. *The Transparency of Spectacle: Meditations on the Moving Image*. Albany: State University of New York Press, 1998.

Dmytryk, Edward. *On Film Editing: An Introduction to the Art of Film Construction*. Boston: Focal Press, 1984.

Eisenstein, Sergei. *Film Form: Essays in Film Theory*. Edited and translated by Jay Leyda. New York: Harcourt Brace Jovanovich, 1949.

Eisenstein, Sergei. *Selected Works, Vol. I, Writings, 1922–34*. Edited and translated by Richard Taylor. London: BFI; Bloomington and Indianapolis: Indiana University Press, 1988.

Eisenstein, Sergei. *Selected Works, Vol. II, Towards a Theory of Montage*. Edited by Michael Glenny and Richard Taylor, translated by Michael Glenny. London: BFI, 1991.

Elliott, David. *New Worlds: Russian Art and Society, 1900-1937*. New York: Rizzoli, 1986.

Elsaesser, Thomas. "The New Hollywood: Cinema Beyond Distance and Proximity." In *Moving Images, Culture and the Mind*, edited by I. Bondebjerg, 187–203. Luton: University of Luton Press, 2000.

Elsaesser, Thomas. "Afterword – Digital Cinema and the Apparatus: Archaeologies, Epistemologies, Ontologies." In *Cinema and Technology: Cultures, Theories, Practices*, edited by Bruce Bennett, Marc Furstenau, and Adrian Mackenzie, 226–40. Basingstoke and New York: Palgrave Macmillan, 2008.

Elsaesser, Thomas and Kay Hoffmann, eds., *Cinema Futures: Cain, Abel, or Cable? Cinema Futures in the Digital Age*. Amsterdam: Amsterdam University press, 1998.

Elster, Jon. *Ulysses Unbound: Studies in Rationality, Precommitment, and Restraint*. Cambridge: Cambridge University Press, 2000.

Enticknap, Leo. *Moving Image Technology: From Zoetrope to Digital*. London: Wallflower Press, 2005.

Fischer, Lucy. "Film Editing." In *A Companion to Film Theory*, edited by Toby Miller and Robert Stam, 64–83. Malden, MA and Oxford: Blackwell, 1999.

Friedberg, Anne. "The End of Cinema: Multimedia and Technological Change." In *Reinventing Film Studies*, edited by Christine Gledhill and Linda Williams, 438–52. London: Arnold, 2000.

Furstenau, Marc, ed. *The Film Theory Reader: Debates and Arguments*. London and New York: Routledge, 2010.

Furstenau, Marc. "Film Editing, Digital Montage, and the 'Ontology' of Cinema," *Cinémas* 28, nos. 2–3 (Spring 2018): 29–49.

Furstenau, Marc. "Review of Angela Dalle Vacche, *André Bazin's Film Theory*," *Canadian Journal of Film Studies* 30, no. 2 (Fall 2021): 116–20.

Furstenau, Marc. "Realism, Illusion, and Special Effects in the Cinema." In Lefebvre and Furstenau, *Special Effects on the Screen*, 131–63.

Gaudreault, André and Laurent Le Forestier. *De l'assemblage au montage cinématographique: Instauration et standardisation d'une pratique*. Montréal: Les Presses de l'Université de Montréal, 2022.

Gaut, Berys. "Digital Cinema." In *The Routledge Companion to Philosophy and Film*. Edited by Paisley Livingstone and Carl Plantinga, 75–85. Abingdon: Routledge, 2009.

Gaut, Berys. *A Philosophy of Cinematic Art*. Cambridge: Cambridge University Press, 2010.

Gaut, Berys and Paisley Livingston, eds. *The Creation of Art: New Essays in Philosophical Aesthetics*. Cambridge: Cambridge University Press, 2003.

Genette, Gérard. *The Work of Art: Immanence and Transcendence*, translated by G. M. Goshgarian. Ithaca: Cornell University Press, 1997.

Genette, Gérard. *The Aesthetic Relation*. Translated by G. M. Goshgarian. Ithaca: Cornell University Press, 1999.

Genette, Gérard. "From Text to Work." In *Essays in Aesthetics*, translated by Dorrit Cohn, 1–28. Lincoln: University of Nebraska Press, 2005.

Goodman, Nelson. *Languages of Art: An Approach to a Theory of Symbols*. Indianapolis: Hackett, 1976.

Grusin, Richard and Jocelyn Szczepaniuk-Gillece, eds. *Ends of Cinema*. Minneapolis: University of Minnesota Press, 2020.

Hanson, Matt. *The End of Celluloid: Film Futures in the Digital Age*. Mies: Rotovision, 2004.

Haugeland, John. "Analog and Analog." *Philosophical Topics* 12 (1981): 213–26.

Hauser, Arnold. *The Social History of Art, Volume 4: Naturalism, Impressionism, the Film Age*. New York: Vintage, 1958.

Heath, Stephen. "The Cinematic Apparatus: Technology as Historical and Cultural Form." In De Lauretis and Heath, *The Cinematic Apparatus*, 1–13.

Higgins, Scott. "The Silent Screen, 1895–1927: Editing." In Keil and Whissel, *Editing and Special/Visual Effects*, 22–36.

Hill, John and Pamela Church Gibson, eds. *The Oxford Guide to Film Studies*. Oxford: Oxford University Press, 1998.

Hirsch, E. D. *Validity in Interpretation*. New Haven, CT: Yale University Press, 1967.

Hyman, John. *The Objective Eye: Color, Form, and Reality in the Theory of Art*. Chicago: University of Chicago Press, 2006.

Irwin, William. *Intentionalist Interpretation: A Philosophical Explanation and Defense*. Westport, CT: Greenwood Press, 1999.

Ivasović, Sergej. "Digital Editing," *Moveast*, 7 (2002): 151–57.

Jarvie, Ian. *Philosophy of the Film: Epistemology, Ontology, Aesthetics*. New York and London: Routledge and Kegan Paul, 1987.

Johnson, William. "*Russian Ark*," *Film Quarterly* 57, no. 2 (Winter 2003–2004): 48–51.

Katz, Matthew. "Analog and Digital Representation." *Minds and Machines* 18 (2008): 403–8.

Keane, Stephen. *CineTech: Film, Convergence, and New Media*. Basingstoke: Palgrave Macmillan, 2007.

Keil, Charlie and Kristin Whissel, eds. *Editing and Special/Visual Effects*. New Brunswick, NJ: Rutgers University Press, 2016.

Kindem, Gorham A. "Peirce's Semiotic Phenomenalism and Film," *Quarterly Review of Film Studies* 4, no. 1 (Winter, 1979): 61–69.

Kipnis, Laura. "Film and Changing Technologies." In Hill and Church Gibson, *The Oxford Guide to Film Studies*, 595–604.

Kolker, Robert. "The Film Text and Film Form." In Hill and Church Gibson, *The Oxford Guide to Film Studies*, 11–29.

Kracauer, Siegfried. *Theory of Film: The Redemption of Physical Reality*. Princeton: Princeton University Press, 1960.

Kujundzic, Dragan. "After 'After': The *Ark*ive Fever of Alexander Sokurov," *Quarterly Review of Film and Video* 21, no. 3 (2004): 219–39.

Kuleshov, Lev. *Kuleshov on Film: Writings of Lev Kuleshov*. Translated and edited by Ronald Levaco. Berkeley, Los Angeles: University of California Press, 1974.

Langer, Suzanne K. *Feeling and Form: A Theory of Art*. New York: Charles Scribner's Sons, 1953.

Lefebvre, Martin and Marc Furstenau. "Digital Editing and Montage: The Vanishing Celluloid and Beyond," *Cinémas* 13, nos. 1–2 (Fall 2002): 69–107.

Lefebvre, Martin and Marc Furstenau, eds. *Special Effects on the Screen: Faking the View from Méliès to Motion Capture*. Amsterdam: Amsterdam University Press, 2022.

Lewis, David. "Analog and Digital," *Noûs* 5, no. 3 (1971): 321–27.

Leyda, Jay. *Kino: A History of the Russian and Soviet Film*. Princeton: Princeton University Press, 1960.

LoBrutto, Vincent. *Selected Takes: Film Editors on Editing.* New York and Westport, CT: Praeger, 1991.

MacInnes, Paul. "How We Got the Shot: Cary Fukunaga on *True Detective*'s Tracking Shot," *The Guardian*, March 17, 2014, https://www.theguardian.com/tv-and-radio/tvandradioblog/2014/mar/17/true-detective-cary-fukunaga-tracking-shot.

Maley, Corey. "Analog and Digital, Continuous and Discrete." *Philosophical Studies* 155, no. 1 (2011): 117–31.

Malraux, André. "Sketch for a Psychology of the Moving Pictures." In *Reflections on Art: A Source Book of Writings by Artists, Critics, and Philosophers*, edited by Suzanne K. Langer, 317–27. Baltimore: The Johns Hopkins Press, 1958.

Manovich, Lev. *The Language of New Media.* Cambridge, MA: MIT Press, 2001.

Margolis, Joseph. "The Ontological Peculiarity of Works of Art." In *Art and Philosophy*, 17–24. Brighton: Harvester Press, 1980.

Margolis, Joseph. "Mechanical Reproduction and Cinematic Humanism." In *What, After All, Is a Work of Art?*, 101–28. University Park, PA: The Pennsylvania State University Press, 1999.

Maynard, Patrick. *The Engine of Visualization: Thinking Through Photography.* Ithaca and London: Cornell University Press, 1997.

Metz, Christian. "The Cinema: Language or Language System?" In *Film Language: A Semiotics of the Cinema*, translated by Michael Taylor, 31–91. New York: Oxford University Press, 1974.

Metz, Christian. *Impersonal Enunciation, or The Place of Film.* Translated by Cormac Deane. New York: Columbia University Press, 2016.

Mitchell, William J. *The Reconfigured Eye: Visual Truth in the Post-Photographic Era.* Cambridge, MA: MIT Press, 1994.

Mitry, Jean. *Esthétique et psychologie du cinéma I: Les Structures.* Paris: Éditions Universitaires, 1963.

Mitry, Jean. "Entretien intempestif: L'histoire en question," *Cinématographe* 110 (May 1985): 76–79.

Mitry, Jean. *The Aesthetics and Psychology of the Cinema.* Translated by Christopher King. Bloomington and Indianapolis: Indiana University Press, 1997.

Monticone, Paul. "Classical Hollywood, 1928–1946: Editing." In Keil and Whissel, *Editing and Special/Visual Effects*, 51–67.

Morgan, Daniel. "Rethinking Bazin: Ontology and Realist Aesthetics," *Critical Inquiry* 32 (Spring 2006): 443–81.

Mulvey, Laura. *Death 24x a Second.* London: Reaktion, 2006.

Münsterberg, Hugo. *The Photoplay: A Psychological Study.* New York and London: D. Appleton and Co., 1916.

Murch, Walter. *In the Blink of an Eye: A Perspective on Film Editing.* Los Angeles: Silman-James Press, 1995.

Ohanian, Thomas. *Digital Nonlinear Editing: New Approaches to Editing Film and Video*. Boston and London: Focal Press, 1993.

Orpen, Valerie. *Film Editing: The Art of the Expressive*. London: Wallflower Press, 2003.

Oudart, Jean-Pierre. "La Suture, I et II," *Cahiers du cinéma*, 211, 212 (April–May 1969): 36–39; 50–56.

Overpick, Deron. "The New Hollywood, 1981–1999: Editing." In Keil and Whissel, *Editing and Special/Visual Effects*, 129–41.

Penfold, Christopher. "Elizaveta Svilova and Soviet Documentary Film." PhD diss., University of Southampton, 2013. https://eprints.soton.ac.uk/367302/1/C%2520Penfold%2520Thesis.pdf.

Peirce, Charles S. *Collected Papers of Charles Sanders Peirce*, Vol. 8. Edited by Arthur W. Burk. Cambridge, MA: Harvard University Press, 1958.

Perkins, V. F. *Film as Film: Understanding and Judging Movies*. Harmondsworth: Penguin, 1972.

Perl, Jed. *Authority and Freedom: A Defense of the Arts*. New York: Alfred A. Knopf, 2021.

Polan, Dana. "Postwar Hollywood, 1947–1967: Editing." In Keil and Whissel, *Editing and Special/Visual Effects*, 78–90.

Prall, D. W. *Aesthetic Judgment*. New York: Thomas Y. Crowell Co., 1929.

Prince, Stephen. "True Lies: Perceptual Realism, Digital Images, and Film Theory," *Film Quarterly* 49, no. 3 (Spring, 1996): 27–37.

Prince, Stephen. *A New Pot of Gold: Hollywood Under the Electronic Rainbow,1980–1989*. History of the American Cinema, edited by Charles Harpole, vol. 10. New York: Charles Scribner's Sons, 2000.

Prince, Stephen *Digital Visual Effects in Cinema: The Seduction of Reality*. New Brunswick, NJ: Rutgers University Press, 2012.

Pudovkin, V. I. *Film Technique and Film Acting*. Translated by Ivor Montagu. London: Vision Press, 1954.

Purse, Lisa "Working Space: *Gravity* and the Digital Long Take." In *The Long Take: Critical Approaches*, edited by John Gibs and Douglas Pye, 221–38. London: Palgrave Macmillan, 2017.

Ravetto-Biagioli, Kriss. "Floating on the Borders of Europe: Sokurov's *Russian Ark*," *Film Quarterly* 59, no. 1 (Fall 2005): 17–22.

Reisz, Karel with Gavin Millar. *The Technique of Film Editing*, 2nd ed. New York and London: Focal Press/Hastings House, 1968.

Ridley, Aaron. "Expression in Art." In *The Oxford Handbook of Aesthetics*, edited by Jerrold Levinson, 211–27. Oxford: Oxford University Press, 2003.

Rizov, Vadim. "The 29 Features Shot, in Whole or in Part, on 35mm in U.S. Release Year 2023," *Filmmaker* 32, no. 2 (Winter 2024): 120–25.

Robert Friedel, *A Culture of Improvement: Technology and the Western Millennium*. Cambridge, MA and London: MIT Press, 2007.

Robin, Régine. *Socialist Realism: An Impossible Aesthetic*. Stanford: Stanford University Press, 1992.

Rodowick, D. N. *The Virtual Life of Film*. Cambridge, MA: Harvard University Press, 2007.

Rohdie, Sam. *Montage*. Manchester and New York: Manchester University Press, 2006.

Rombes, Nicholas. *Cinema in the Digital Age*. London: Wallflower Press, 2009.

Rosen, Philip. *Change Mummified: Cinema, Historicity, Theory*. Minneapolis and London: University of Minnesota Press, 2001.

Rubin, Michael. *Nonlinear: A Field Guide to Digital Video and Film Editing*, 4th ed. Gainesville, FL: Triad Publishing, 2000.

Rubin, Michael. *Droidmaker: George Lucas and the Digital Revolution*. Gainesville, FL: Triad Publishing, 2006.

Salt, Barry. *Film Style and Technology: History and Analysis*. London: Starword, 1982.

Schaeffer, Jean-Marie. *Art of the Modern Age: Philosophy of Art from Kant to Heidegger*. Translated by Steven Rendall. Princeton: Princeton University Press, 2000.

Schenk, Sonja. *The Digital Filmmaking Handbook*, 4th ed. Boston: Course Technology, 2012.

Scruton, Roger. "Photography and Representation." In *The Aesthetic Understanding: Essays in the Philosophy of Art and Culture*, 102–26. Manchester: Carcanet Press, 1983.

Searle, John. *Intentionality: An Essay in the Philosophy of Mind*. Cambridge: Cambridge University Press, 1983.

Searle, John. *Seeing Things as They Are: A Theory of Perception*. Oxford: Oxford University Press, 2015.

Shaw, Jeffrey and Peter Wiebel, eds., *Future Cinema: The Cinematic Imaginary After Film*. Cambridge, MA: MIT Press, 2003.

Singer, Irving. *Reality Transformed: Film as Meaning and Technique* (Cambridge, MA: MIT Press, 1998).

Souriau, Étienne. *La correspondance des arts: Éléments d'esthétique comparée*. Paris: Flammarion, 1947.

Sparshott, F. E. "Basic Film Aesthetics." *Journal of Aesthetic Education* 5, no. 2 (April 1971): 11–34.

Spottiswoode, Raymond. *Film and Its Techniques*. Berkeley and Los Angeles: University of California Press, 1969.

Stam, Robert. *Film Theory: An Introduction*. Malden, MA: Blackwell, 2000.

Stewart, Garrett. *Framed Time: Toward a Postfilmic Cinema*. Chicago: University of Chicago Press, 2007.

Tarkovsky, Andrei. *Sculpting in Time: Reflections on the Cinema.* Translated by Kitty Hunter-Blair. Austin: University of Texas Press, 1986.

Theisen, Eric. "The Story of the Moviola," *The International Photographer* 7, no. 10 (November 1935): 4–5.

Thomson-Jones, Katherine. *Image in the Making: Digital Innovation and the Visual Arts.* Oxford: Oxford University Press, 2021.

Tsivian, Yuri. "Montage Theory I (Hollywood Continuity)." In Branigan and Buckland, *Routledge Encyclopedia of Film Theory*, 306–13.

Tsivian, "Montage Theory II (Soviet Avant-Garde)." In Branigan and Buckland, *Routledge Encyclopedia of Film Theory*, 314–21.

Turvey, Malcolm. *Doubting Vision: Film and the Revelationist Tradition.* Oxford: Oxford University Press, 2008.

Willis, Holly. *New Digital Cinema.* London: Wallflower Press, 2005.

Wilson, George. *Narration in Light: Studies in Cinematic Point of View.* Baltimore and London: The Johns Hopkins University Press, 1986.

Wimsatt, W. K. and M. C. Beardsley, "The Intentional Fallacy," *The Sewanee Review* 54, no. 3 (July–September 1946): 468–88.

Wollen, Peter. *Signs and Meaning in the Cinema,* enlarged edition. Bloomington: Indiana University Press, 1972.

Wollen, Peter. "Cinema and Technology: A Historical Overview." In *The Cinematic Apparatus*, edited by Teresa De Lauretis and Stephen Heath, 14–22. London: Macmillan, 1980.

Wollheim, Richard. *Art and Its Objects*, 2nd edition. Cambridge: Cambridge University Press, 1980.

Wood, Aylish. *Digital Encounters.* Abingdon: Routledge, 2007.

Wright, Benjamin. "The Auteur Renaissance, 1968–1980: Editing." In Keil and Whissel, *Editing and Special/Visual Effects*, 103–15.

Yacavone, Daniel. *Film Worlds: A Philosophical Aesthetics of Cinema.* New York: Columbia University Press, 2015.

Index

For Product Safety Concerns and Information please contact our EU representative GPSR@taylorandfrancis.com
Taylor & Francis Verlag GmbH, Kaufingerstraße 24, 80331 München, Germany

www.ingramcontent.com/pod-product-compliance
Ingram Content Group UK Ltd.
Pitfield, Milton Keynes, MK11 3LW, UK
UKHW020015200926
475702UK00014B/1891
9781041187134